Tiny Machine Learning Techniques for Constrained Devices

Tiny Machine Learning Techniques for Constrained Devices explores the cutting-edge field of Tiny Machine Learning (TinyML), enabling intelligent machine learning on highly resource-limited devices such as microcontrollers and edge Internet of Things (IoT) nodes. This book provides a comprehensive guide to designing, optimizing, securing, and applying TinyML models in real-world constrained environments.

This book offers thorough coverage of key topics, including:

- Foundations and Optimization of TinyML: Covers microcontroller-centric power optimization, core principles, and algorithms essential for deploying efficient machine learning models on embedded systems with strict resource constraints.
- Applications of TinyML in Healthcare and IoT: Presents innovative use cases such as compact artificial intelligence (AI) solutions for healthcare challenges, real-time detection systems, and integration with low-power IoT and low-power wide-area network (LPWAN) technologies.
- Security and Privacy in TinyML: Addresses the unique challenges of securing TinyML deployments, including privacy-preserving techniques, blockchain integration for secure IoT applications, and methods for protecting resource-constrained devices.
- Emerging Trends and Future Directions: Explores the evolving landscape of TinyML research, highlighting new applications, adaptive frameworks, and promising avenues for future investigation.
- Practical Implementation and Case Studies: Offers hands-on insights and real-world examples demonstrating TinyML in action across diverse scenarios, providing guidance for engineers, researchers, and students.

This book is an essential resource for embedded system designers, AI practitioners, cybersecurity professionals, and academics who want to harness the power of TinyML for smarter, more efficient, and secure edge intelligence solutions.

Tiny Machine Learning Techniques for Constrained Devices

Edited by
Khalid El Makkaoui, Ismail Lamaakal,
Ibrahim Ouahbi, Yassine Maleh, and
Ahmed A. Abd El-Latif

CRC Press
Taylor & Francis Group
Boca Raton London New York

CRC Press is an imprint of the
Taylor & Francis Group, an **informa** business

A CHAPMAN & HALL BOOK

Designed cover image: Shutterstock

First edition published 2026
by CRC Press
2385 NW Executive Center Drive, Suite 320, Boca Raton FL 33431

and by CRC Press
4 Park Square, Milton Park, Abingdon, Oxon, OX14 4RN

CRC Press is an imprint of Taylor & Francis Group, LLC

© 2026 selection and editorial matter, Khalid El Makkaoui, Ismail Lamaakal, Ibrahim Ouahbi, Yassine Maleh, and Ahmed A. Abd El-Latif; individual chapters, the contributors

ISBN: 978-1-032-89752-3 (hbk)
ISBN: 978-1-032-89753-0 (pbk)
ISBN: 978-1-003-54444-9 (ebk)

DOI: 10.1201/9781003544449

Typeset in Times
by codeMantra

Contents

Preface

In today's interconnected world, intelligence is moving closer to where data is generated: on tiny, resource-limited devices such as sensors, microcontrollers, and wearables. This shift is made possible by Tiny Machine Learning (TinyML), a cutting-edge discipline focused on optimizing machine learning algorithms to run efficiently on devices with severe constraints in power, memory, and computing resources.

Tiny Machine Learning Techniques for Constrained Devices explores this exciting frontier, revealing how artificial intelligence (AI) can be embedded in the smallest devices to enable smarter healthcare, smarter cities, and smarter industries. This book dives into the fundamental principles behind TinyML, the innovations that make it feasible on constrained hardware, and the broad applications reshaping our digital landscape.

From optimizing energy consumption in microcontrollers to deploying neural networks on edge devices, the chapters guide readers through core concepts and practical techniques. You'll discover how TinyML powers real-time applications such as mask detection during pandemics and low-power IoT networks optimized for long-range communication protocols like LoRaWAN. Crucially, this book also tackles the security and privacy challenges inherent to TinyML deployments, including the integration of blockchain to safeguard sensitive data.

The content is structured as follows:

- Foundational techniques for power optimization and algorithm design tailored to constrained hardware
- Real-world applications in healthcare, industrial IoT, and environmental monitoring
- Innovative strategies for communication and energy management in IoT networks
- Security frameworks ensuring privacy and resilience in TinyML-enabled systems
- Exploration of future directions that promise to expand the capabilities of TinyML further

With contributions from leading researchers and practitioners, this volume offers both theoretical grounding and actionable insights. Whether you are an academic, engineer, or enthusiast eager to bring AI into the smallest devices, this book equips you with the knowledge and tools to design and implement effective TinyML solutions.

Are you ready to embark on a journey into the realm where machine learning meets minimalism? Together, let's unlock the power of TinyML and shape the next generation of intelligent, connected devices.

About the Editors

Prof. Khalid El Makkaoui received his Master's degree in Networks and Systems and a Ph.D. in Computer Science from Hassan 1st University, Settat, Morocco, in 2014 and 2018. Since 2019, he has been an Associate Professor with the Department of Computer Science at the Multidisciplinary Faculty of Nador, University Mohammed Premier, Oujda, Morocco. His research interests focus on cybersecurity and artificial intelligence. He has published over 40 papers (book chapters, international journals, and conferences).

Dr. Ismail Lamaakal (Member, IEEE) received a Master of Science degree in Computer Science from the Multidisciplinary Faculty of Nador, University Mohammed Premier, Oujda, Morocco. He is currently advancing toward his Ph.D. in Computer Science at the same esteemed institution. As an Artificial Intelligence Scientist, his research primarily focuses on the innovative integration of Tiny Machine Learning, the Internet of Things (IoT), Human–computer interaction, and Embedded Systems. His work is characterized by its pioneering approach in the field, emphasizing practical applications and advancements in these interconnected domains. His contributions are marked by a commitment to pushing the boundaries of AI and its applications in the modern technological landscape.

Prof. Ibrahim Ouahbi is Professor of Computer Science at the Multidisciplinary Faculty of Nador, University Mohammed Premier, Oujda, Morocco. He received his Ph.D. in Didactics of Informatics in 2018 from Sidi Mohammed Ben Abdellah University, Fez, Morocco. He was Professor of Educational Technologies at the Faculty of Educational Sciences, Mohammed V University of Rabat, in 2019. His research interests include artificial intelligence, cybersecurity, and ICT integration in science education and learning.

Prof. Yassine Maleh (http://orcid.org/0000-0003-4704-5364) is a professor of cyber-security and computer sciences at Sultan Moulay Slimane University, since 2019. He is a Senior Member of IEEE. Dr. Maleh has made contributions in the fields of Information Security and Privacy, Internet of Things Security, and Wireless and Constrained Networks Security. His research interests include Information Security and Privacy, Internet of Things, Network Security, Information system, and IT Governance. He has published more than 200 papers (book chapters, international journals, and conferences/workshops), 50 edited books, and 7 authored book. He is the Editor-in-Chief of the *International Journal of Smart Security Technologies* (IJSST). He serves as an Associate Editor for *IEEE Access* (2019 Impact Factor 4.098), the *International Journal of Digital Crime and Forensics* (IJDCF), and the *International Journal of Information Security and Privacy* (IJISP). He was also a Guest Editor of a special issue on Recent Advances on Cyber Security and Privacy for Cloud-of-Things of the *International Journal of Digital Crime and Forensics*

(IJDCF), Volume 10, Issue 3, July–September 2019. He has served and continues to serve on executive and technical program committees and as a reviewer of numerous international conferences and journals such as *Elsevier Ad Hoc Networks, IEEE Network Magazine, IEEE Sensor Journal, ICT Express*, and *Springer Cluster Computing*. He was the Publicity Chair of BCCA 2019 and the General Chair of the MLBDACP 19 symposium.

Prof. Ahmed A. Abd El-Latif (Senior Member, IEEE) is a Professor at Menoufia University, Egypt, and holds academic positions at Prince Sultan University, Saudi Arabia. He earned his Ph.D. (2013) from the Harbin Institute of Technology, China. With extensive research leadership, Dr. Abd El-Latif has spearheaded and contributed to numerous national and international research projects. He has authored over 350 publications in prestigious journals and conferences, and has received multiple awards for his contributions.

Since 2022, he has served as the Head of the MEGANET 6G Lab Research in the Russian Federation. Additionally, he holds key leadership roles, including Vice Chair of the EIAS Research Lab and Founder and Deputy Director of the Center of Excellence in Quantum and Intelligent Computing at Prince Sultan University, Saudi Arabia.

An active contributor to the scientific community, Dr. Abd El-Latif serves as Editor-in-Chief for several journals, Chair/Co-Chair of international conferences, and Series Book Editor. His research spans quantum communications and cryptography, cybersecurity, AIoT (artificial intelligence of things), AI-based image processing, information hiding, and applications of dynamical systems (e.g., chaotic systems and quantum walks) in cybersecurity.

Contributors

Qasem Abu Al-Haija
Department of Cybersecurity
Jordan University of Science and
    Technology (JUST)
Irbid, Jordan

Mohammed R. Al-Matari
Department of Computer Information
    Systems
Jordan University of Science and
    Technology (JUST)
Irbid, Jordan

Mouncef Filali Bouami
Department of Computer Science
Multidisciplinary Faculty of Nador
University Mohammed Premier
Oujda, Morocco

Khalid El Makkaoui
Department of Computer Science
Multidisciplinary Faculty of Nador
University Mohammed Premier
Oujda, Morocco

Siham Essahraui
Department of Computer Science
Multidisciplinary Faculty of Nador
University Mohammed Premier
Oujda, Morocco

Ismail Lamaakal
Department of Computer Science
Multidisciplinary Faculty of Nador
University Mohammed Premier
Oujda, Morocco

Yassine Maleh
Laboratory LaSTI, Department of
    Networks, Telecommunications and
    Cybersecurity
ENSA Khouribga, Sultan Moulay
    Slimane University
Beni Mellal, Morocco

Doaa Omar
Department of Network Engineering
    and Security
Jordan University of Science and
    Technology (JUST)
Irbid, Jordan

Ibrahim Ouahbi
Department of Computer Science
Multidisciplinary Faculty of Nador
University Mohammed Premier
Oujda, Morocco

Chaymae Yahyati
Department of Computer Science
Multidisciplinary Faculty of Nador
University Mohammed Premier
Oujda, Morocco

1 Microcontroller-Centric Power Optimization in Embedded Systems

Siham Essahraui, Yassine Maleh,
Khalid El Makkaoui, Mouncef Filali Bouami,
and Ibrahim Ouahbi

1.1 INTRODUCTION

The ongoing advancement of embedded systems has profoundly reshaped modern technology, empowering a wide array of devices—from everyday consumer appliances to sophisticated industrial automation systems—with intelligent capabilities [1]. Embedded systems, by nature, are seamlessly integrated into larger mechanical or electronic frameworks and are engineered to deliver reliable, efficient, and often real-time functionality. As technologies such as the Internet of Things (IoT) [2], edge computing, and smart environments continue to evolve, there is a growing expectation for embedded systems to maintain high performance while minimizing energy usage [3].

Against this backdrop, the development of low-power embedded systems has become a focal point for innovation and research [4]. Energy efficiency is especially crucial in scenarios involving portable, battery-powered, or remote devices, where energy limitations directly impact device longevity, usability, and environmental sustainability. Achieving this balance requires careful optimization of both hardware components—such as microcontrollers [5], sensors, and wireless modules—and software elements, including firmware, power-aware scheduling, and task management [6].

This chapter aims to provide a comprehensive overview of methodologies for developing energy-efficient embedded systems. It examines microcontroller architectures and power-saving mechanisms at the component level, highlights effective hardware selection practices, and discusses software techniques to reduce power consumption. The discussion also incorporates real-world applications and comparative analysis of development platforms, offering practical insights for engineers and researchers alike.

The structure of this chapter is as follows: Section 1.2 outlines the fundamental principles and key applications of embedded systems. Section 1.3 explores microcontroller design and architecture-level strategies for power optimization. Section 1.4 addresses the selection of low-power components. Section 1.5 details

DOI: 10.1201/9781003544449-1

software-oriented approaches and low-power hardware platforms, while Section 1.6 concludes with a discussion of emerging trends and future directions in embedded system development.

1.2 OVERVIEW OF EMBEDDED SYSTEMS AND THEIR APPLICATIONS

Embedded systems are dedicated computing platforms engineered to perform specific tasks within a broader mechanical or electronic environment [7]. In contrast to general-purpose computers, these systems are designed with a focus on high efficiency, robust reliability, and often real-time operational capability. Their presence is widespread, underpinning a vast array of technologies such as consumer devices, automated industrial machinery, healthcare instrumentation, and automotive electronics.

1.2.1 APPLICATIONS ACROSS SECTORS

As illustrated in Figure 1.1, embedded systems are integral to numerous sectors:

- **Automotive:** In modern vehicles, embedded systems are essential for functions such as engine management, advanced driver assistance systems (ADAS), airbag control, and infotainment systems, contributing to enhanced safety, efficiency, and user experience [8,9].
- **Medical Devices:** Within healthcare, embedded technology forms the basis of critical devices like pacemakers, infusion systems, and diagnostic instruments, enabling real-time data acquisition and precise therapeutic control to support patient health [10].
- **Internet of Things:** These systems serve as the computational core of IoT devices, powering smart appliances, fitness wearables, and industrial monitoring systems that transmit real-time data for automation and analytics [11–13].
- **Industrial Automation:** Embedded systems streamline manufacturing operations by enabling robotics, automated process control, and predictive maintenance, thereby enhancing productivity and reducing operational expenses [14].

The domain of embedded systems is rapidly advancing to meet new technological challenges. Among current priorities, low-power design stands out, especially for devices that depend on limited power sources such as batteries. Advanced methods like dynamic voltage scaling, power gating, and the implementation of optimized sleep modes are key strategies aimed at reducing energy consumption without degrading system performance. These innovations are especially critical in IoT, wearable computing, and remote sensing, where operational longevity and energy efficiency are paramount.

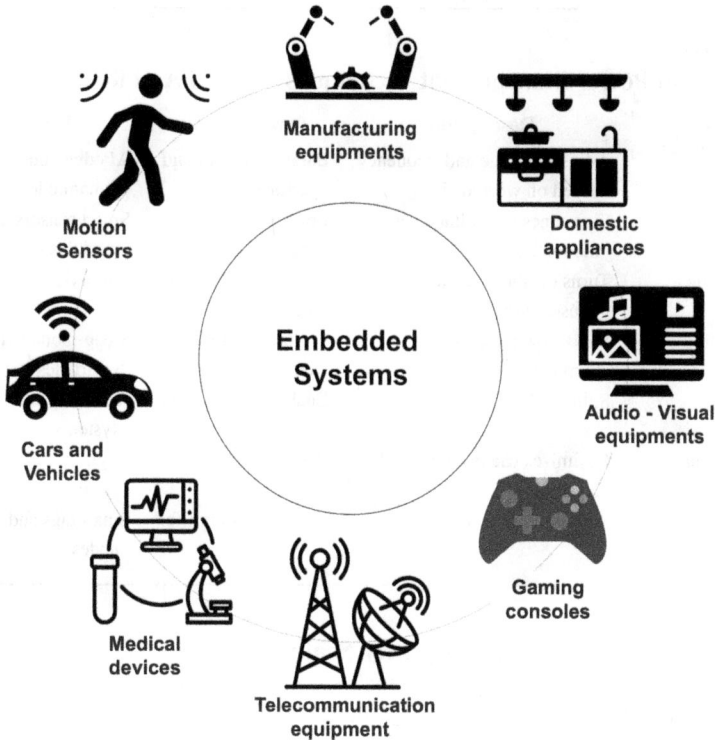

FIGURE 1.1 Applications of embedded systems in modern technology.

This chapter delves into the principles, design techniques, and development tools employed in building low-power embedded systems, with an emphasis on facilitating intelligent, energy-efficient processing at the edge [15,16].

1.3 MICROCONTROLLER ARCHITECTURES AND POWER OPTIMIZATION TECHNIQUES

Modern microcontroller unit (MCU) architectures are essential to developing low-power embedded systems [2]. As the demand for energy efficiency continues to grow—particularly in edge computing and IoT environments—designers need to apply both hardware- and software-level strategies. This section explores core power-saving methods integrated into MCU designs and examines the latest trends in ultra-low-power (ULP) systems [17,18].

Table 1.1 provides a clear overview of widely used MCU power management techniques at the architectural level, outlining how they work, their advantages, and where they're typically applied. This structured comparison facilitates informed decision-making during embedded system development.

TABLE 1.1

Summary of Power Management Techniques in MCU Architectures

Technique	Description	Power Benefit	Use Case
DVFS	Adjusts voltage and frequency based on workload.	Balances power and performance.	AI edge nodes with dynamic load
AVS	Fine-tunes the voltage per module.	Lowers per-block energy use.	Smart sensors and IoT modules
Clock gating	Turns off the clock to unused parts.	Cuts dynamic power loss.	Battery-powered MCUs
Power gating	Shuts down idle blocks completely.	Minimizes static leakage.	Sleep-mode IoT and wearables
Pipelining	Splits tasks into stages.	Enables lower clock speed.	Real-time control systems
Cache Mgmt.	Optimizes cache config and access.	Reduces memory power use.	Logging and buffer systems
Low-power modes	Switches MCU to sleep states.	Extends battery life.	Smart tags and sensor nodes

- **Dynamic Voltage and Frequency Scaling (DVFS)** enables a system to modify its supply voltage and clock speed on the fly based on workload demands. When activity is low or the system is idle, it can lower these settings to conserve energy. Conversely, under heavier loads, the system boosts voltage and frequency to maintain speed. DVFS is a vital tool for balancing power use, performance, and heat management. It often works alongside other techniques like power and clock gating, as long as the MCU supports dynamic adjustments [19].
- **Adaptive Voltage Scaling (AVS) and Clock/Power Gating** fine-tunes the voltage provided to specific components in real time, reacting to changing conditions such as temperature. This helps boost energy efficiency while preserving functionality. Clock gating disables the clock signal to inactive components, cutting down on dynamic power draw. Power gating goes further by turning off power to unused sections, reducing static power and leakage currents. These methods are crucial for detailed and responsive power management in embedded designs.
- **Number of Cores and Core Pipelining:** Using multiple cores spreads tasks across several low-power processors instead of pushing a single high-speed core. This setup lowers both power density and heat output. Pipelining improves performance by dividing instruction execution into steps that can run simultaneously. Thoughtful use of pipelining and controlled parallelism helps avoid wasteful computations and keeps energy use down.
- **Memory** is often a major power drain in embedded systems. Flash memory offers non-volatile storage and is typically accessible without activating the CPU. SRAM is used for caching because it's fast and energy-efficient. DRAM—especially asynchronous types—needs frequent refreshing, making it less ideal for systems that sleep often. Choosing the right cache size,

associativity, and write policy, along with compiler tricks like loop unrolling, can help reduce memory-related power consumption [20].

- **Power Down/Saving Modes** Low-power MCUs come with several modes that selectively shut down certain functions. These modes differ in how quickly they wake up and how much current they use. Using them effectively requires knowing how peripherals behave during sleep and how fast the system needs to be ready again. Designers must ensure that the MCU not only supports these modes but also moves between active and sleep states efficiently [21].
- **New Trends** is reshaping how embedded systems save power. Power management integrated circuits (PMICs) now offer fine-grained control over power delivery and timing. "Dark silicon" strategies deactivate unused chip areas to stop leaks. Modern heterogeneous CPU architectures architectures match tasks to either powerful or energy-saving cores based on urgency. The open and flexible RISC-V architecture is gaining momentum. Lastly, harvesting energy from the environment—like sunlight or vibrations—is a promising alternative to batteries in some devices [22].

1.4 COMPONENT SELECTION FOR LOW-POWER EMBEDDED SYSTEMS

The development of energy-conscious embedded systems mandates a comprehensive strategy in hardware selection. While optimizing the central microcontroller is essential, equal emphasis must be placed on evaluating the performance and power characteristics of peripherals, sensors, and communication interfaces. Each hardware component contributes cumulatively to the system's overall power consumption, a factor of critical importance in battery-operated and edge AI deployments. This section outlines methodological approaches for identifying components that achieve a judicious trade-off between functional performance and minimal energy expenditure (see Table 1.2).

1.4.1 MCU SELECTION

At the core of every embedded architecture lies the MCU, whose careful selection is pivotal for realizing low-power objectives. Assessment criteria for MCUs include current consumption in active and idle modes, shutdown current, integrated system-on-chip (SoC) features, and the availability of low-power operational states such as sleep and deep sleep. ULP MCUs typically exhibit active-mode consumption in the range of 30–40 μA/MHz, with shutdown current potentially as low as 50–70 nA. However, real-world power efficiency extends beyond nominal datasheet specifications. It is imperative for system designers to evaluate configuration flexibility, access to evaluation boards, and the comprehensiveness of vendor-provided development ecosystems. Effective energy management further necessitates the selective activation of internal modules and the utilization of software-based energy profiling tools to detect and mitigate high-consumption operations.

TABLE 1.2

Component Selection Strategies for Low-Power Embedded Systems

Component	Selection Criteria	Low-Power Strategies
Microcontroller unit (MCU)	• Low supply and shutdown current • SoC integration and sleep modes	• Use ULP MCUs (30–40 µA/MHz) • Enable only essential peripherals • Use profiling tools for power analysis
Analog-to-digital converter (ADC)	• Architecture: SAR vs. Sigma-Delta • Sampling resolution and rate	• Prefer SAR ADCs for efficiency • Use burst-mode sampling • Integrate Class AB op-amps
Digital-to-analog converter (DAC)	• Output buffering and impedance • Update rate requirements	• Use sample-and-hold techniques • Minimize output frequency • Allow DAC to operate in sleep modes
Digital interfaces	• Protocol type (UART, SPI, I²C) • Clocking and idle-state handling	• Use event-triggered communication • Reduce clock speed and duty cycle • Utilize DMA for data transfers
Wireless interfaces	• Stack complexity • Energy per transmission	• Transmit in short bursts • Use lightweight stacks (e.g., BLE) • Employ low-duty-cycle operation
Sensors	• Accuracy and power profile • Ease of integration	• Use MEMS sensors with edge ML • Select models with low quiescent current • Minimize sensing frequency where possible

1.4.2 PERIPHERAL SELECTION

Peripherals—whether embedded within the chip or externally interfaced—exert significant influence on the system's energy dynamics. Their selection should be guided by a detailed examination of current consumption during various operational states and the presence of mechanisms enabling dynamic mode control.

1.4.2.1 ADC Selection

Analog-to-digital converters (ADCs) facilitate the acquisition of analog signals and are indispensable in sensor-integrated systems. ADC architectures, predominantly successive approximation register (SAR) and sigma-delta, differ in their energy implications. SAR ADCs are often favored in ULP contexts due to their lower energy requirements, albeit with potential compromises in resolution. Additional strategies for minimizing power include reducing sampling rates, employing burst-mode acquisition, and selecting operational amplifiers that utilize Class AB topology over the less efficient Class A. Integration of digital filtering within ADC modules further enhances energy-efficient signal processing, particularly for low-amplitude inputs.

1.4.2.2 DAC Selection

While digital-to-analog converters (DACs) are less frequently deployed in typical IoT applications compared to ADCs, their optimization remains essential in relevant use cases. Key considerations include buffer configuration, update frequency, and load impedance. Techniques such as sample-and-hold circuitry, output duty cycling, and the deployment of direct memory access (DMA) for data transfer can substantially lower energy consumption. Some low-power DAC designs offer persistent analog output functionality even during deep sleep phases, thereby enabling intermittent signal generation without engaging the primary processor.

1.4.2.3 Interfaces

Digital communication interfaces—namely UART, SPI, and I²C—are vital for peripheral integration yet represent a considerable share of energy expenditure. Interface selection should prioritize protocol efficiency, support for power-saving idle states, and compatibility with low-duty-cycle operation modes. Event-driven communication schemes typically outperform polling mechanisms in terms of energy efficiency. Additional power-saving measures include dynamic clock adjustment and DMA-based data handling, which are particularly advantageous in scenarios where high-frequency communication is non-essential.

1.4.3 SENSOR SELECTION

Sensors act as the primary conduits for physical data acquisition, and their energy profiles directly influence embedded system efficiency. Selection criteria must encompass not only functional parameters such as sensitivity, resolution, and operational range but also power consumption in both active and quiescent modes. Preference should be given to compact sensors with low static current draw and seamless integration capabilities with MCUs. Modern MEMS-based sensors provide a compelling combination of precision and low power usage and often incorporate embedded signal processing features that reduce reliance on continuous MCU engagement. Moreover, sensors equipped with TinyML processing enable on-node inference, substantially reducing data transmission requirements and extending battery life. Ultimately, informed sensor selection contributes to both hardware optimization and streamlined software architecture, yielding a more efficient embedded solution.

1.5 EMBEDDED SYSTEMS DESIGN AND OPTIMIZATION

The growing need for portable, intelligent, and autonomous technologies has significantly accelerated the development of energy-efficient embedded platforms. As embedded systems become increasingly prevalent across a broad spectrum of applications, reducing power consumption—while maintaining high levels of performance and reliability—has emerged as a primary design goal. This section examines the two fundamental aspects of embedded system optimization: software-level efficiency techniques and the deployment of low-power hardware solutions [23]. Together, these elements serve as the cornerstone for creating sustainable and high-performing embedded solutions in today's technology landscape.

1.5.1 SOFTWARE OPTIMIZATION TECHNIQUES FOR LOW-POWER SYSTEMS

The role of embedded software in achieving energy-efficient operation is paramount. Strategic firmware design, intelligent event-handling, thoughtful GPIO configuration, integration of real-time operating systems, and the use of diagnostics tools all contribute significantly to optimizing power usage during both active and standby states of a device.

- **Firmware Optimization Techniques:** Efficient firmware serves as the backbone of power-aware system design. Techniques such as extending low-power mode durations, deactivating superfluous peripherals, and fine-tuning power and clock domains are fundamental. Additionally, implementing proper shutdown routines to place components in their minimal energy state can markedly enhance system longevity and reduce energy overhead.
- **Interrupt vs. Polling Paradigms:** Constant polling by the CPU can lead to unnecessary power consumption. Adopting an interrupt-driven model allows the system to remain dormant until an event necessitates processing, thereby conserving energy. This approach is especially effective for sporadic operations like sensor readings or wireless transmissions.
- **GPIO and Peripheral Management:** Precision in managing GPIO states and peripheral interfaces is essential. Configuring unused pins to avoid floating states, disabling extraneous internal pull resistors, and employing low-power modes for peripherals contribute to minimizing both static and dynamic energy loss. Disabling clock signals to inactive units is another best practice for efficient power use.
- **Real-Time Operating Systems (RTOS):** RTOS solutions provide fine-grained control over task execution and system idling. They enable energy-saving features such as tickless idle, prioritized task handling, and sleep-state transitions. Leveraging these capabilities ensures that the system performs necessary functions while remaining in low-power states whenever feasible.
- **Code Profiling and Static Analysis Tools:** Energy-aware development benefits greatly from visibility into code behavior. Profiling tools help uncover power-intensive routines, while static analysis can identify inefficiencies such as memory leaks and suboptimal loops. These insights guide developers toward refining their codebase for performance and energy conservation.

1.5.2 LOW-POWER HARDWARE PLATFORMS

To accelerate the development of energy-efficient embedded systems, a range of hardware platforms are available. These systems feature microcontrollers, integrated sensors, and communication modules engineered for minimal power draw and are ideal for applications in IoT, TinyML, and edge computing (see Table 1.3).

TABLE 1.3

Comparison of Low-Power Hardware Platforms and Feature Support

Platform	BLE	Dual Core	AI/TFLite	Low Sleep Current <2 µA	Open Source	RTOS	Secure Boot
Arduino Portenta H7 (Cortex-M7 + M4)	✓	✓	✓	✓	✓	✓	✓
Raspberry Pi RP2040 (Dual Cortex-M0+)	✓	✓	×	✓	✓	✓	×
Jetson Nano (NVIDIA Tegra GPU)	×	✓	✓	×	✓	✓	✓
STM32U5 (Cortex-M33)	✓	×	×	✓	✓	✓	✓
CC2651R3 (Cortex-M4)	✓	×	×	✓	✓	✓	✓
MAX32670 (Cortex-M4)	✓	×	×	✓	✓	✓	✓
EFR32MG24 (Cortex-M33)	✓	×	×	✓	✓	✓	✓
CY8C4247LQQ (Cortex-M0)	✓	×	×	✓	✓	✓	✓
ATSAML21E (Cortex-M0+)	✓	×	×	✓	✓	✓	✓
LPC55S66 (Cortex-M33)	✓	✓	×	✓	✓	✓	✓

- **Arduino:** Arduino's low-power solutions include the Nano 33 BLE Sense with a Cortex-M4 core and embedded motion sensors. The Portenta H7, equipped with dual Cortex-M7 and M4 cores, enables simultaneous real-time and high-level processing. Both boards support open-source firmware development and TensorFlow Lite integration.
- **Raspberry Pi:** The RP2040 microcontroller, based on dual Cortex-M0+ cores, delivers substantial performance within stringent energy constraints. The Pico board is suitable for sensor-based applications, while the Raspberry Pi 4 is geared toward higher-performance workloads with higher power needs.
- **Jetson Nano:** Designed by NVIDIA, Jetson Nano is a compact AI platform featuring GPU acceleration. While it draws more power than traditional MCUs, it strikes a balance for edge AI applications, offering efficient vision processing and Linux compatibility.
- **ST Microelectronics:** The STM32L0, L4, and U5 MCU families provide ULP operation across Cortex-M0+ to M4 cores. The IoT Discovery Kit integrates wireless and sensor modules, making it an ideal prototyping tool for connected, power-sensitive systems.
- **Texas Instruments:** TI's wireless MCUs—CC13xx, CC23xx, and CC26xx—are engineered for minimal energy usage. The CC2651R3 supports BLE, Zigbee, and WPAN, with an active current as low as 2.9 mA and a shutdown current of 0.1 µA, all accessible via the SimpleLink SDK.
- **Analog Devices:** The MAX32670, built around a Cortex-M4 core, is designed for rugged industrial IoT. It includes hardware-based AES and CRC, efficient voltage regulation, and support for standard communication protocols, maintaining performance with low energy demands.
- **Silicon Labs:** The EFR32MG24 and BG22 are ULP SoCs ideal for BLE and wireless IoT. These devices feature sleep-mode currents as low as 1.2 µA and are backed by robust development kits and complete firmware stacks.
- **Infineon Technologies:** The CY8C4247LQQ-BL483 integrates a Cortex-M0 MCU with BLE, capacitive touch, and a 12-bit SAR ADC. It supports low-power operation with an active current at 1.7 mA and a sleep current of 1.5 µA, making it well suited for wearable and interactive applications.
- **Microchip Technology:** Featuring advanced power control, the ATSAML21E offers current levels of just 35 µA/MHz when active and 200 nA during sleep. Its backup battery mode is ideal for applications requiring prolonged low-energy operation.
- **NXP Semiconductors:** The LPC55S66 includes a 150-MHz Cortex-M33 processor, secure boot, and generous memory with optimized power performance. Its versatile I/O and communication support make it suitable for secure AI-driven systems at the edge.

1.6 CONCLUSION AND FUTURE WORKS

Embedded systems are becoming increasingly indispensable across a diverse array of domains, including industrial automation, medical technology, smart consumer devices, and IoT-based infrastructure. With rising demands for both energy efficiency and performance, integrating low-power design methodologies at the hardware and software levels has become a central theme in embedded system engineering. This chapter has explored the key strategies for reducing energy consumption, such as optimizing microcontroller architecture, selecting energy-efficient components, and designing power-conscious firmware.

Beyond these established practices, the field is undergoing a significant transformation with the advent of new design paradigms. The integration of artificial intelligence into embedded systems is redefining traditional development approaches. One notable advancement is TinyML, which enables ULP devices to perform real-time, on-device machine learning for applications such as predictive maintenance, activity monitoring, and intelligent sensing. In parallel, the rise of edge computing is encouraging a move away from centralized cloud processing toward localized data analysis, further highlighting the importance of energy-efficient embedded architectures.

Looking ahead, the future of embedded systems will likely include broader adoption of RISC-V-based microcontrollers, providing open and customizable hardware platforms. Developments in energy harvesting will extend device autonomy, while AI-powered tools for power profiling and system optimization will streamline the design process. Moreover, as embedded systems scale in complexity and connectivity, stronger interoperability standards and security frameworks will be essential.

REFERENCES

1. Lamaakal, I., Maleh, Y., El Makkaoui, K., Ouahbi, I., Pławiak, P., Alfarraj, O., … & Abd El-Latif, A. A. (2025). Tiny language models for automation and control: Overview, potential applications, and future research directions. *Sensors, 25* (5), 1318.
2. Lamaakal, I., Essahraui, S., Maleh, Y., El Makkaoui, K., Ouahbi, I., Bouami, M. F., … & Niyato, D. (2025). A comprehensive survey on tiny machine learning for human behavior analysis. *IEEE Internet of Things Journal, 12* (6), 32419–32443.
3. Lamaakal, I., Ouahbi, I., El Makkaoui, K., Maleh, Y., Pławiak, P., & Alblehai, F. (2024). A TinyDL model for gesture-based air handwriting of Arabic numbers and simple Arabic letter recognition. *IEEE Access, 12*, 76589–76605.
4. Lamaakal, I., El Makkaoui, K., Ouahbi, I., & Maleh, Y. (2024). A TinyML model for gesture-based air handwriting of Arabic number recognition. *Procedia Computer Science, 236*, 589–596.
5. Lamaakal, I., El Mourabit, N., El Makkaoui, K., Ouahbi, I., & Maleh, Y. (2024, October). Efficient gesture-based recognition of tifinagh characters in air handwriting with a TinyDL model. In *2024 Sixth International Conference on Intelligent Computing in Data Sciences (ICDS)*, Marrakech, Morocco (pp. 1–8). IEEE.

6. Lamaakal, I., Maleh, Y., Ouahbi, I., El Makkaoui, K., & Abd El-Latif, A. A. (2024, May). A Deep Learning-Powered TinyML Model for Gesture-Based Air Handwriting of Simple Arabic Letter Recognition. In *International Conference on Digital Technologies and Applications*, Benguerir, Morocco (pp. 32–42). Springer.

7. Subero, A. (2021). Embedded systems overview. In *Programming Microcontrollers with Python: Experience the Power of Embedded Python*, Subero, Armstrong (pp. 77–105). Springer.

8. Ashjaei, M., Lo Bello, L., Daneshtalab, M., Patti, G., Saponara, S., & Mubeen, S. (2021). Time-sensitive networking in automotive embedded systems: State of the art and research opportunities. *Journal of Systems Architecture*, *117*, 102137. https://doi.org/10.1016/j.sysarc.2021.102137.

9. Essahraui, S., Lamaakal, I., El Hamly, I., Maleh, Y., Ouahbi, I., El Makkaoui, K., Filali Bouami, M., Pławiak, P., Alfarraj, O., & Abd El-Latif, A. A. (2025). Real-time driver drowsiness detection using facial analysis and machine learning techniques. *Sensors*, *25* (3), 812. https://doi.org/10.3390/s25030812.

10. Arandia, N., Garate, J. I., & Mabe, J. (2022). Embedded sensor systems in medical devices: Requisites and challenges ahead. *Sensors*, *22* (24), 9917. https://doi.org/10.3390/s22249917.

11. Essahraui, S., Ouahbi, I., El Makkaoui, K., & Filali Bouami, M. (2024). A deep learning-driven fingerprint verification model for enhancing exam integrity in Moroccan higher education. *Information Security Journal: A Global Perspective*, *33*, 1–13. https://doi.org/10.1080/19393555.2024.2374247.

12. Eceiza, M., Flores, J. L., & Iturbe, M. (2021). Fuzzing the internet of things: A review on the techniques and challenges for efficient vulnerability discovery in embedded systems. *IEEE Internet of Things Journal*, *8* (13), 10390–10411. https://doi.org/10.1109/JIOT.2021.3056179.

13. Essahraui, S., Bouyardan, M., El Hamly, I., El Makkaoui, K., Ouahbi, I., & Filali Bouami, M. (2024). Enhancing exam integrity in Moroccan higher education: An AI-based fingerprint verification model. In *Proceedings of the 7th International Conference on Networking, Intelligent Systems and Security*, Meknes, Morocco (pp. 1–5). https://doi.org/10.1145/3659677.3659720.

14. Shylla, D., Shah, P., Sekhar, R., & others. (2023). Embedded systems in industrial automation 4.0. In *2023 14th International Conference on Computing Communication and Networking Technologies (ICCCNT)*, Delhi, India (pp. 1–6). IEEE. https://doi.org/10.1109/ICCCNT56998.2023.10307700.

15. Ajani, T. S., Imoize, A. L., & Atayero, A. A. (2021). An overview of machine learning within embedded and mobile devices–optimizations and applications. *Sensors*, *21* (13), 4412.

16. Essahraui, S., El Makkaoui, K., Filali Bouami, M., & Ouahbi, I. (2024). CNN and Vision Transformer models for detecting cheating in online examinations: A comparative evaluation. In *The International Conference on Artificial Intelligence and Smart Environment*, Errachidia, Morocco (pp. 295–301). Springer.

17. Warden, P., & Situnayake, D. (2019). *TinyML: Machine Learning with TensorFlow Lite on Arduino and Ultra-Low-Power Microcontrollers*. O'Reilly Media.

18. Benini, L., & de Micheli, G. (2000). System-level power optimization: Techniques and tools. *ACM Transactions on Design Automation of Electronic Systems (TODAES)*, *5* (2), 115–192. https://doi.org/10.1145/335043.335044.

19. Murray, J., Lu, T., Pande, P., & Shirazi, B. (2013). Sustainable DVFS-enabled multicore architectures with on-chip wireless links. In M. Zelkowitz (Ed.), *Advances in Computers* (Vol. 88, pp. 125–158). Elsevier. https://doi.org/10.1016/B978-0-12-407725-6.00003-4.

20. Chen, J., Loi, I., Flamand, E., Tagliavini, G., Benini, L., & Rossi, D. (2023). Scalable hierarchical instruction cache for ultralow-power processor clusters. *IEEE Transactions on Very Large Scale Integration (VLSI) Systems*, *31* (4), 456–469. https://doi.org/10.1109/TVLSI.2022.3228336.
21. Jadhav, S., & Chaudhari, B. S. (2024). Embedded systems for low-power applications. In B. S. Chaudhari, S. N. Ghorpade, M. Zennaro, & R. Paškauskas (Eds.), *TinyML for Edge Intelligence in IoT and LPWAN Networks* (pp. 13–26). Elsevier.
22. Han, H., & Siebert, J. (2022). TinyML: A systematic review and synthesis of existing research. In the *2022 International Conference on Artificial Intelligence in Information and Communication (ICAIIC)*, Jeju Island, Republic of Korea (pp. 269–274). https://doi.org/10.1109/ICAIIC54071.2022.9722636.
23. White, E. (2024). *Making Embedded Systems: Design Patterns for Great Software*. O'Reilly Media, Inc.

2 Core Principles and Algorithms for Tiny Machine Learning

*Siham Essahraui, Yassine Maleh,
Khalid El Makkaoui, and
Mouncef Filali Bouami*

2.1 INTRODUCTION

The rapid evolution of Internet of Things (IoT) [1,2] technologies and the increasing proliferation of embedded systems have created a pressing demand for intelligent, energy-efficient, and real-time data processing at the edge of the network. Conventional machine learning approaches often rely on centralized cloud infrastructures for both training and inference [3]. However, this reliance introduces several limitations, including high communication latency, data privacy risks, and substantial energy consumption.

To address these challenges, the integration of machine learning with resource-constrained embedded devices—referred to as Tiny Machine Learning (TinyML) [4]—has emerged as a transformative paradigm in edge computing. TinyML enables the deployment of compact and optimized machine learning models directly onto ultra-low-power microcontrollers and edge devices. This capability empowers real-time decision-making, reduces dependence on cloud infrastructure, and strengthens data privacy through on-device inference. The synergy between TinyML and edge AI is paving the way for innovative applications across diverse domains, including healthcare, precision agriculture, industrial automation, and intelligent transportation systems [5,6].

The primary objective of this chapter is to explore the fundamental principles, architectural components, and algorithmic innovations that define TinyML, with an emphasis on its practical deployment within edge environments. Additionally, this chapter investigates how artificial intelligence (AI) techniques can be leveraged to optimize key edge computing functions such as task offloading, energy efficiency, resource allocation, and system security.

This chapter is structured as follows. Section 2.2 introduces the foundational concepts and architectural framework of TinyML. Section 2.3 outlines the end-to-end workflow for developing and deploying TinyML models. Section 2.4 examines the convergence of AI and edge computing, presenting various paradigms and algorithmic categories applicable in edge scenarios. Section 2.5 presents a detailed analysis

DOI: 10.1201/9781003544449-2

of AI-enabled strategies for optimizing offloading, energy consumption, security, privacy, and resource management. Finally, Section 2.6 concludes the chapter with a summary of key findings and insights into emerging research directions in TinyML and edge AI.

2.2 FOUNDATIONS OF TinyML

Serving as a foundational element for intelligent edge systems, TinyML represents the convergence of machine learning and embedded computing, specifically tailored for environments with severe resource limitations. This section outlines the core principles and objectives of TinyML, highlighting its role in enabling localized intelligence, facilitating real-time decision-making, and augmenting existing cloud and edge computing infrastructures.

2.2.1 PRINCIPLES AND OBJECTIVES

TinyML [7,8] supports the integration of lightweight machine learning models into compact edge devices that typically operate under stringent resource constraints, including limited computational capabilities, restricted memory capacity, and low power availability. By enabling local data processing and interpretation, TinyML allows these devices to perform real-time responses without the need for constant cloud connectivity. Rather than replacing fog or cloud computing paradigms, TinyML serves as a complementary technology that enhances and accelerates existing computational frameworks.

The principal aim of TinyML is to improve the operational efficiency of machine learning models by minimizing computational complexity and reducing the volume of data required. This capability is pivotal to advancing edge AI technologies and driving the expansion of the IoT [9]. Within this framework, TinyML involves the development and deployment of specialized architectures, frameworks, methodologies, tools, and strategies that enable on-device analytics across a range of sensor modalities—including visual, auditory, speech, motion, chemical, physical, textual, and cognitive inputs. These applications are specifically designed to operate within extremely low power budgets, typically in the milliwatt range or below, making them ideal for battery-powered embedded systems. The technology's primary focus is large-scale deployment in domains such as IoT infrastructures and wireless sensor networks.

2.2.2 ARCHITECTURAL OVERVIEW

As depicted in Figure 2.1, the architecture of TinyML functions at the convergence of cloud computing and edge computing paradigms. Cloud platforms are responsible for executing computationally intensive machine learning tasks using high-performance processors such as tensor processing units (TPUs), graphics processing units (GPUs), and central processing units (CPUs). These systems are ideal for large-scale model training and batch inference. In contrast, edge computing enables the deployment of machine learning models on localized infrastructure—such as servers, routers, and

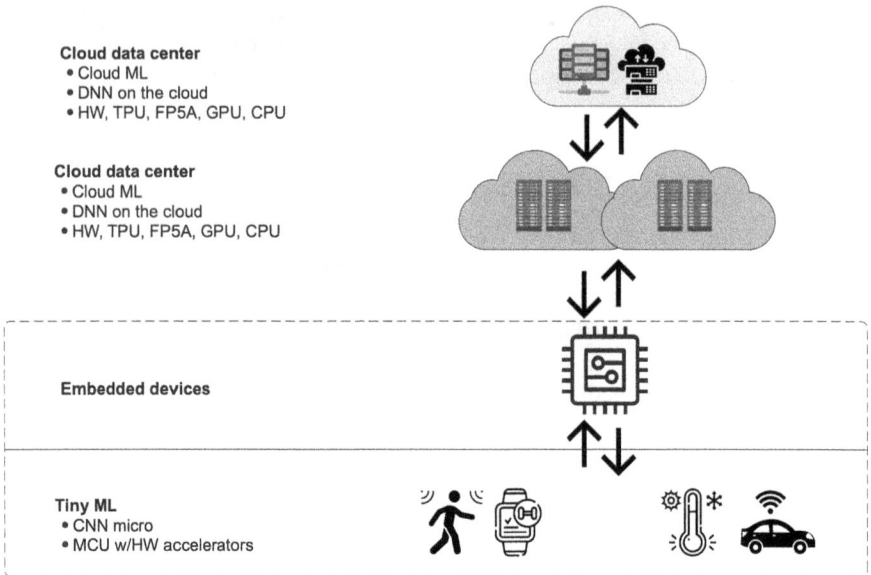

FIGURE 2.1   Cloud computing, edge computing, and TinyML architecture [7].

switches—thus situating processing closer to the data source. This proximity significantly reduces latency, lessens dependency on cloud resources, and enhances real-time responsiveness [10]. TinyML advances this model by facilitating inference directly on microcontrollers and embedded edge devices. This architectural innovation significantly reduces both bandwidth consumption and latency, while also enhancing privacy and autonomy in IoT environments. By supporting on-device inference, TinyML ensures system responsiveness even in settings with limited or intermittent connectivity [3].

Deep learning constitutes a foundational component of TinyML, empowering complex functionalities such as image classification and natural language processing on resource-constrained platforms [11]. However, the computational intensity of deep neural networks (DNNs) necessitates specialized hardware for efficient operation. Examples of such hardware include:

- **Tensor Processing Units (TPUs):** Custom-designed accelerators that enable energy-efficient, low-latency inference of deep learning models.
- **FP5A Processing Units:** Flexible processors capable of functioning at various bit-widths to optimize computational efficiency in embedded systems.
- **CNN Microarchitectures:** Lightweight convolutional neural networks, such as CNN Micro, tailored for operation within stringent power and memory limitations.

These hardware optimizations, in tandem with lightweight model architectures, facilitate the deployment of advanced AI functionalities in embedded systems, thereby accelerating innovation in edge AI.

The integration of cloud computing, edge computing, and TinyML results in a synergistic framework that fortifies the IoT landscape. Each tier of this architecture delivers distinct yet complementary capabilities:

- **Cloud Computing:** Offers scalable storage and computational resources ideal for high-level data analytics and model training.
- **Edge Computing:** Brings computation closer to the data origin to minimize latency and facilitate real-time analytics.
- **TinyML:** Provides ultra-low-power, on-device intelligence, enabling smart, autonomous behavior in embedded devices.

This hierarchical integration enhances the efficiency of data processing, decreases bandwidth usage, and supports a wide range of advanced applications, including predictive maintenance, smart home automation, and health monitoring systems. Together, these layers constitute a robust and intelligent foundation for next-generation IoT solutions [7].

2.2.3 Core Components: Hardware, Software, and Algorithms

TinyML signifies a transformative approach to deploying machine learning models directly onto ultra-low-power embedded systems. This innovation enables localized data processing at the network edge, thereby reducing latency, minimizing energy consumption, and mitigating privacy concerns. The underlying architecture of a TinyML system hinges on the cohesive integration of three core components:

- **Hardware:** IoT and edge devices that incorporate microcontrollers specifically engineered for energy-efficient operation and capable of performing local machine learning inference.
- **Software:** Lightweight operating systems, cloud connectivity interfaces, and communication protocols that facilitate the deployment, maintenance, and management of ML models on embedded platforms.
- **Algorithms:** Machine learning models optimized for environments with limited resources, often utilizing techniques such as quantization, pruning, and model compression to ensure computational efficiency.

As illustrated in Figure 2.2, these components are co-developed to support efficient on-device computation, preserve data fidelity, and promote seamless integration with practical IoT applications. Achieving high-performance in TinyML systems necessitates a meticulous hardware-software co-design strategy. This involves aligning compact, efficient software frameworks with highly optimized ML models while maintaining robust standards for data security and integrity [12].

In real-world deployment scenarios, TinyML models are generally compiled into binary formats and uploaded—or "flashed"—onto microcontroller units (MCUs) from a more capable host system, thereby completing the deployment workflow.

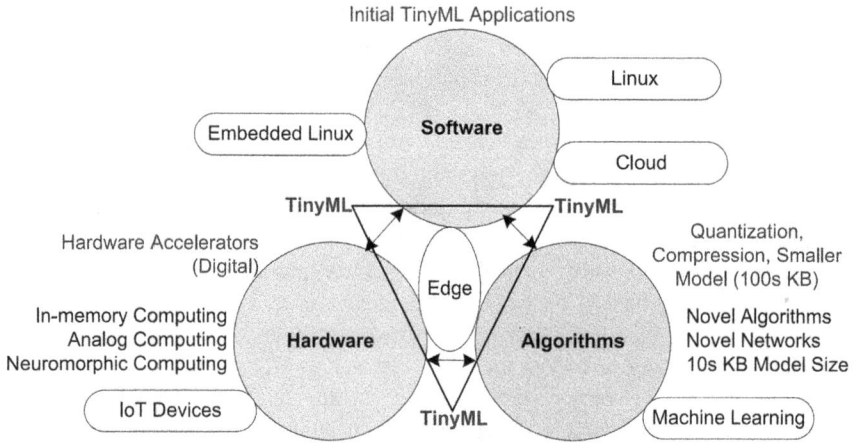

FIGURE 2.2   TinyML components: hardware, software, and algorithms [12].

2.2.4   KEY BENEFITS FOR TINYML

The incorporation of TinyML into edge computing infrastructures introduces a range of benefits that significantly improve the performance, scalability, and security of IoT systems (see Table 2.1). Key advantages include:

- **Energy Efficiency:** TinyML models executed on MCUs exhibit substantially lower energy consumption compared to conventional processors and GPUs.

     This high efficiency is particularly advantageous for battery-powered or energy-harvesting devices, facilitating their use in remote or mobile environments. Furthermore, their integration into battery-supported platforms such as electric scooters or mobile robots enables prolonged operational durations.

- **Cost-Effective Deployment:** Many IoT applications necessitate significant computational resources, often driving reliance on expensive cloud services. TinyML mitigates this dependence by performing local processing on affordable microcontrollers. These MCUs typically operate within a frequency range of 1–400 MHz, offering memory capacities between 2 and 512 kB, and storage ranging from 32 KB to 2 MB. Their low production cost and operational simplicity make them ideal candidates for large-scale, distributed deployments.

- **Latency Reduction:** Local inference execution through TinyML eliminates the need to transmit data to distant cloud servers, thereby minimizing latency and enhancing real-time responsiveness. This attribute is critical in time-sensitive or mission-critical scenarios. Additionally, local computation alleviates pressure on cloud infrastructure, resulting in more responsive and robust systems overall.

TABLE 2.1

Comparison of TinyML, Edge Computing, and Traditional Cloud-Based ML for Edge Intelligence

Feature/Benefit	TinyML	Edge Computing	Cloud-Based ML
Energy efficiency	✓	✓	✗
Low-cost implementation	✓	✓	✗
Reduced latency	✓	✓	✗
Real-time inference	✓	✓	✗
Offline operation	✓	✗	✗
System reliability	✓	✓	✗
Data privacy	✓	✓	✗
Scalable model training	✗	✗	✓
Compute resource availability	✗	✓	✓
Model complexity support	✗	✓	✓

- **Improved Reliability and Data Privacy:** Conventional IoT architectures rely on transmitting raw data to centralized servers for processing, exposing the system to vulnerabilities such as transmission errors and cyber threats. In contrast, TinyML performs data analysis directly on-device, which decreases the volume and sensitivity of data being transmitted. This localized approach enhances system reliability and offers a higher degree of privacy protection.

2.3 WORKFLOW OF TinyML MODELS

Embedded systems form the foundational computational layer of TinyML, characterized by their ultra-low power requirements and minimal hardware capacity. Typically, these platforms offer only a few kilobytes to several hundred kilobytes of RAM, under 2 MB of flash storage, and operate on less than 3 mW of power consumption. In comparison, even the most energy-efficient mobile devices consume several hundred milliwatts to multiple watts, highlighting the superior energy efficiency of MCUs in TinyML deployments. This efficiency makes MCUs particularly well suited for executing machine learning tasks in highly resource-constrained environments [8].

The development cycle for TinyML generally encompasses three core phases: training, optimization, and deployment. While the training process mirrors conventional ML workflows, the optimization and deployment stages are uniquely adapted to the limitations of embedded systems (see Figure 2.3).

- **Training Stage:** This initial phase involves selecting a suitable machine learning algorithm and collecting training datasets, which can originate from public repositories or directly from embedded sensors. Model training

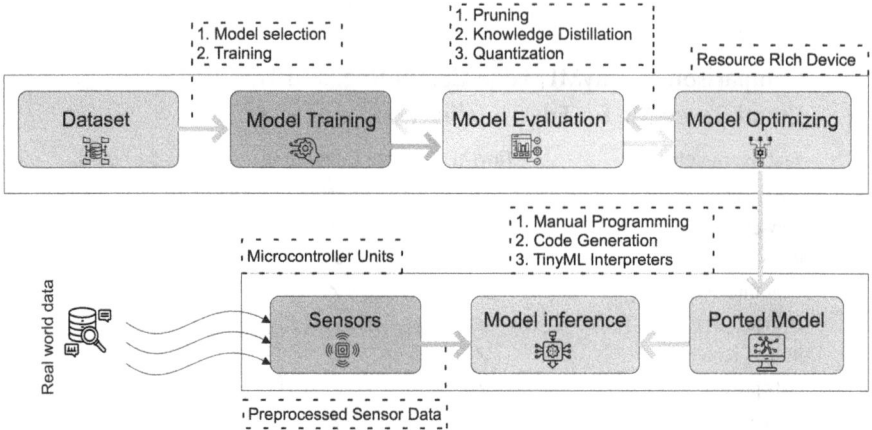

FIGURE 2.3 The TinyML workflow from data collection to model deployment.

is typically conducted on high-performance computing platforms—such as workstations or servers—using established ML frameworks like TensorFlow, PyTorch, or Scikit-learn. The outcome is a high-accuracy, but computationally intensive model.

- **Optimization Stage:** To ensure compatibility with embedded devices, the trained model undergoes a series of optimization procedures:
  - *Pruning:* Eliminates non-essential weights and neurons, thereby reducing model size and complexity while preserving performance.
  - *Knowledge Distillation (KD):* Transfers knowledge from a complex "teacher" model to a smaller "student" model to achieve compression with retained accuracy.
  - *Quantization:* Converts high-precision computations (e.g., 32-bit floats) into lower-precision representations (e.g., 8-bit integers), substantially decreasing memory usage and accelerating inference.

  These methods must be carefully balanced to maintain the integrity and predictive capabilities of the model [13].

- **Deployment Stage:** Once optimized, the model is converted into a hardware-executable format—typically as a C or C++ array—using tools such as TensorFlow Lite for Microcontrollers (TFLM). The compiled model is then embedded directly into the device's firmware.

  Post-deployment, the MCU performs real-time inference on incoming sensor data. Because all computations occur locally, TinyML enables ultra-low-latency responses independent of cloud connectivity. This localized intelligence makes TinyML a pivotal enabler of edge AI across sectors such as healthcare, transportation, smart infrastructure, and education.

2.4 THE CONVERGENCE OF AI AND EDGE COMPUTING

The fusion of AI with edge computing marks a pivotal transformation in the delivery of intelligent services across the IoT ecosystem. Edge computing processes data in close proximity to its source, significantly reducing latency and improving responsiveness. Concurrently, AI augments system capabilities by enabling real-time analytics and adaptive control mechanisms.

This convergence is primarily motivated by the limitations of cloud-based infrastructures in latency-sensitive and bandwidth-constrained scenarios. The deployment of AI algorithms at the network edge is increasingly common, aiming to optimize computational resources, enhance data privacy, and minimize communication overhead. This integrated approach allows for intelligent data interpretation directly on devices, supporting diverse applications such as autonomous vehicles, smart urban infrastructure, and industrial process automation [14].

2.4.1 PARADIGMS OF AI-ENABLED EDGE COMPUTING

Edge AI architectures can be categorized into distinct paradigms based on the distribution of AI workloads between the cloud and edge environments.

2.4.1.1 Cloud-Assisted Training with Local Inference

In this paradigm, the cloud is utilized for the resource-intensive training of AI models. Once trained, these models are deployed to edge devices, where they perform inference tasks locally. This setup effectively balances high model accuracy with low-latency execution and is particularly suited for applications that demand real-time responsiveness but do not require continuous learning.

2.4.1.2 Full AI Delegation to Edge Devices

In scenarios that demand strict data privacy or operate under limited connectivity, both training and inference processes are conducted entirely on edge devices. Although this approach poses considerable challenges in terms of computational and energy resources, advancements in TinyML, federated learning, and model compression techniques have made it increasingly viable. It enables devices to autonomously adapt to new data and environmental changes.

2.4.2 AI ALGORITHM CATEGORIES IN EDGE CONTEXTS

AI algorithms deployed at the network edge span a wide spectrum—from lightweight classical models to advanced deep learning and reinforcement learning techniques [15]. Each algorithmic category provides distinct advantages that align with the resource constraints and application-specific requirements inherent to edge environments. In such contexts, selecting an appropriate AI model is critical for achieving an optimal balance among performance, latency, energy efficiency, and memory usage.

AI significantly contributes to the advancement of edge computing, which faces considerable optimization challenges due to its inherently distributed architecture and the temporal and spatial variability of workloads across devices. These complexities often result in non-convex optimization problems that scale with the number of users and interconnected nodes. Traditional rule-based systems are often inadequate in addressing such issues. In contrast, AI—particularly machine learning algorithms—can effectively identify latent patterns within the noisy and heterogeneous data streams that are typical of edge settings.

Furthermore, edge computing systems require intelligent adaptability to manage diverse tasks, including computational offloading, resource allocation, energy optimization, latency reduction, and enhancement of user experience. Therefore, the development of AI algorithms for edge applications must consider stringent constraints, such as limited processing power, constrained energy availability, and the need for real-time decision-making.

To address these multifaceted challenges, several categories of AI algorithms are strategically utilized within edge environments.

2.4.2.1 Traditional Machine Learning

Algorithms such as support vector machines (SVM), k-nearest neighbors (KNN) [16], decision trees, and clustering methods are commonly used in edge contexts due to their low computational demands. These models are effective for structured data analysis and are frequently applied in classification, regression, and anomaly detection tasks, making them ideal for real-time, on-device processing.

2.4.2.2 Deep Learning Architectures

Advanced architectures like convolutional neural networks (CNNs) [17], recurrent neural networks (RNNs), and long short-term memory (LSTM) [18] networks are increasingly optimized for edge deployment. These models are particularly adept at processing unstructured data, including images, audio, and video streams. Optimization techniques such as pruning, quantization, and knowledge distillation make deployment feasible on constrained devices, enabling applications in surveillance, speech recognition, and environmental monitoring.

2.4.2.3 Reinforcement and Deep Reinforcement Learning

Reinforcement learning (RL) and its more advanced variant, deep reinforcement learning (DRL), support dynamic decision-making by learning through interaction with the environment. In edge computing, RL is employed for tasks like energy management and task scheduling. Although DRL extends these capabilities to high-dimensional problems, it remains computationally demanding.

2.4.2.4 Federated Learning for Distributed Intelligence

Federated learning (FL) allows multiple edge nodes to collaboratively train a shared global model without transferring raw data. Instead, only model updates are communicated to a central server, preserving user privacy and minimizing bandwidth usage. This decentralized learning model is particularly relevant in sectors with stringent data protection requirements, such as healthcare and finance.

2.4.2.5 Evolutionary Algorithms for Optimization

Inspired by natural evolutionary processes, evolutionary algorithms (EAs) are effective for solving complex optimization problems in distributed systems. Techniques such as genetic algorithms (GA), particle swarm optimization (PSO), and gray wolf optimization (GWO) are used to fine-tune energy consumption, task scheduling, and model configurations. Their robustness and adaptability make them valuable tools in real-time edge AI applications.

2.5 EDGE COMPUTING OPTIMIZATION USING AI SOLUTIONS

This section presents a comprehensive survey of recent research initiatives employing AI techniques to enhance edge computing capabilities. The focus spans several key domains, including task offloading, energy efficiency, system security, data privacy, and resource allocation [19]. Table 2.2 synthesizes the key AI techniques applied across various edge optimization domains, highlighting their strengths and limitations.

2.5.1 COMPUTING OFFLOADING OPTIMIZATION

AI-based task offloading approaches are designed to optimize energy efficiency and minimize latency, with the overarching goal of balancing these two critical performance metrics. Certain methodologies employ deep learning-driven partial offloading mechanisms that dynamically determine the most efficient allocation of tasks. These intelligent models significantly lower energy consumption during operation by tailoring task distribution to the system's current conditions.

To reduce latency, integrated frameworks have been proposed that simultaneously enhance wireless communication, collaborative data caching, and computational efficiency. In addition, heuristic deep learning methods have been developed

TABLE 2.2
AI Techniques for Edge Computing Optimization

Technique	Offloading	Energy Eff.	Security	Privacy	Res. Alloc.
Deep Learning (DL)	✓	✗	✗	✗	✗
Heuristic DL	✓	✗	✗	✗	✗
Q-Learning (RL)	✓	✓	✓	✗	✓
Structural RL	✓	✓	✗	✗	✗
Deep Q-Network (DQN)	✓	✓	✓	✓	✓
Federated Learning (FL)	✓	✗	✗	✓	✓
PDS-learning	✗	✗	✗	✓	✗
Autoencoders/DBNs	✗	✗	✓	✗	✗
Hardware Opt. (FPGA, SRAM)	✗	✓	✗	✗	✗
DRL + MCTS	✗	✗	✗	✗	✓

[20], combining network distance estimation with intelligent search algorithms to determine effective offloading strategies, particularly in environments constrained by limited computational resources.

RL, including classical Q-learning, has also been utilized to support optimal control in edge computing environments. However, traditional RL techniques often struggle with scalability when confronted with high-dimensional state spaces.

In response, researchers have introduced structural RL approaches that incorporate energy awareness and adaptive mechanisms. Some of these methods also integrate energy-harvesting capabilities to reduce dependence on conventional power sources.

To address the scalability issue more effectively, DRL has emerged as a viable solution for handling high-dimensional optimization tasks. Model-free DRL techniques have been successfully applied in complex architectures, including integrated terrestrial–aerial–satellite networks. Variants such as Deep Q-Networks (DQNs) have demonstrated strong performance in approximating optimal policies in dynamic edge environments. Furthermore, combining FL with DRL has proven beneficial for distributed learning scenarios, improving convergence rates while minimizing communication overhead.

2.5.2   Energy Consumption Reduction Using Non-Computation Offloading Methods

Although AI processing on edge devices can raise energy consumption—particularly in time-critical applications like autonomous driving—several strategies have been introduced to optimize power efficiency:

- **Hardware Structure Optimization:** Techniques such as parallel memory access using SRAM and the implementation of binarized DNN accelerators on FPGAs contribute significantly to energy savings. For example, an FPGA-based weed classification system demonstrated a sevenfold energy improvement over GPU-based alternatives [21].
- **Controlling Operating Status:** DRL-based controllers are employed to dynamically manage processor states and communication protocols. These are modeled using Markov decision processes to optimize energy use over mid- and long-term periods.
- **Combining Energy Internet:** The integration of edge computing with renewable energy infrastructures, such as microgrids, facilitates localized energy management. DRL algorithms, often combined with curriculum learning, help balance supply and demand under uncertain production conditions.

2.5.3   Security of Edge Computing

Although edge computing decentralizes data processing—alleviating some cloud-related vulnerabilities—it introduces new security risks, including DDoS and jamming attacks. Classical machine learning models, including minimax Q-learning, stochastic game-based approaches, and ensemble methods, have proven effective in detecting such threats.

Advanced deep learning techniques further improve detection accuracy [5]. For instance, stacked autoencoders combined with softmax classifiers outperform traditional classifiers in identifying malicious behavior. Additionally, models such as deep belief networks and restricted Boltzmann machines have been deployed to detect novel or zero-day attacks.

2.5.4 DATA PRIVACY

While edge computing limits data exposure by reducing cloud transmissions, it remains susceptible to privacy breaches. AI models may inadvertently retain sensitive user data. Counter-measures include:

- **PDS-Learning:** Integrates post-decision state modeling with DQNs to create privacy-aware offloading decisions.
- **Federated Learning:** PAFLM enables collaborative learning across devices without sharing private data.
- **Differential Privacy:** Methods such as Laplace noise injection are applied to model outputs. Frameworks like Edge-Sanitizer combine local differential privacy with deep learning for secure mobile data processing.

2.5.5 RESOURCE ALLOCATION OPTIMIZATION

Dynamic and high-dimensional resource allocation challenges are well suited for DRL techniques. Deep Q-Networks have been deployed to model various system variables, including wireless conditions, trust ratings, caching, and computational resources. Furthermore, self-supervised DNNs trained using Monte Carlo tree search (MCTS) have been shown to reduce system latency by more than 50%.

2.6 CONCLUSION AND FUTURE WORKS

TinyML has established itself as a transformative approach for delivering intelligent, real-time, and energy-efficient analytics at the edge, particularly within environments constrained by limited computational and power resources. This chapter has provided an in-depth exploration of the fundamental concepts, system architectures, algorithmic methodologies, and AI-driven optimization techniques that underpin the design and deployment of TinyML systems.

By focusing on hardware-software co-design, model compression strategies, and edge-specific AI algorithms, this chapter has demonstrated how TinyML effectively addresses the shortcomings of conventional cloud-centric solutions. Localized inference not only minimizes latency and conserves energy but also enhances data privacy and scalability across application domains such as healthcare, agriculture, transportation, and industrial monitoring. Moreover, the integration of AI with edge computing introduces innovative paradigms for task offloading, energy optimization, resource scheduling, and cybersecurity. Intelligent offloading policies and DRL techniques have shown considerable success in adapting to the distributed and dynamic characteristics of edge environments. FL and differential privacy mechanisms have further reinforced secure, decentralized AI infrastructures.

Future research in the field of TinyML and edge AI is oriented toward several critical areas. These include the design and implementation of advanced micro-controllers and neuromorphic hardware architectures optimized for efficient, low-latency on-device inference. Another major direction involves the integration of self-adaptive and continual learning models, enabling edge devices to evolve with changing data and usage contexts. Research efforts are also geared toward enhancing the ability of TinyML systems to perform real-time multimodal fusion across diverse sensor inputs such as vision, speech, and physiological signals. Other priorities encompass the automation of full end-to-end deployment pipelines, from model development and optimization to embedded implementation. Sustainability remains a key concern, with ongoing efforts focused on designing energy-efficient TinyML frameworks that align with environmental goals. Additionally, there is a strong push for the development of standardized evaluation benchmarks and open-source tools to support reproducibility and performance comparison. Finally, improving the robustness, transparency, and security of TinyML systems is essential for their safe integration into mission-critical domains, such as healthcare, autonomous systems, and industrial control environments.

REFERENCES

1. Khanna, A., & Kaur, S. (2019). Evolution of Internet of Things (IoT) and its significant impact in the field of precision agriculture. *Computers and Electronics in Agriculture, 157*, 218–231. https://doi.org/10.1016/j.compag.2018.12.039.
2. Wang, J., Lim, M. K., Wang, C., & Tseng, M.-L. (2021). The evolution of the Internet of Things (IoT) over the past 20 years. *Computers & Industrial Engineering, 155*, 107174. https://doi.org/10.1016/j.cie.2021.107174.
3. Lamaakal, I., Maleh, Y., El Makkaoui, K., Ouahbi, I., Pławiak, P., Alfarraj, O., … & Abd El-Latif, A. A. (2025). Tiny language models for automation and control: Overview, potential applications, and future research directions. *Sensors, 25* (5), 1318.
4. Lamaakal, I., El Makkaoui, K., Ouahbi, I., & Maleh, Y. (2024). A TinyML model for gesture-based air handwriting of Arabic number recognition. *Procedia Computer Science*, 236, 589–596.
5. Essahraui, S., Lamaakal, I., El Hamly, I., Maleh, Y., Ouahbi, I., El Makkaoui, K., Filali Bouami, M., Pławiak, P., Alfarraj, O., & Abd El-Latif, A. A. (2025). Real-time driver drowsiness detection using facial analysis and machine learning techniques. *Sensors, 25* (3), 812. https://doi.org/10.3390/s25030812.
6. Rajapakse, V., Karunanayake, I., & Ahmed, N. (2023). Intelligence at the extreme edge: A survey on reformable TinyML. *ACM Computing Surveys, 55* (13s), 1–30. https://doi.org/10.1145/3583683.
7. Elhanashi, A., Dini, P., Saponara, S., & Zheng, Q. (2024). Advancements in TinyML: Applications, limitations, and impact on IoT devices. *Electronics, 13* (17), 3562. https://doi.org/10.3390/electronics13173562.
8. Lamaakal, I., Essahraui, S., Maleh, Y., El Makkaoui, K., Ouahbi, I., Bouami, M. F., … & Niyato, D. (2025). A comprehensive survey on tiny machine learning for human behavior analysis. *IEEE Internet of Things Journal, 12* (16), 32419–32443.
9. Schizas, N., Karras, A., Karras, C., & Sioutas, S. (2022). TinyML for ultra-low power AI and large-scale IoT deployments: A systematic review. *Future Internet, 14* (12), 363. https://doi.org/10.3390/fi14120363.

10. Dutta, L., & Bharali, S. (2021). TinyML meets IoT: A comprehensive survey. *Internet of Things*, *16*, 100461. https://doi.org/10.1016/j.iot.2021.100461.

11. Lamaakal, I., Maleh, Y., Ouahbi, I., El Makkaoui, K., & Abd El-Latif, A. A. (2024, May). A deep learning-powered TinyML model for gesture-based air handwriting simple arabic letters recognition. In *International Conference on Digital Technologies and Applications* (pp. 32–42), Benguerir, Morocco. Springer Nature Switzerland.

12. Ray, P. P. (2022). A review on TinyML: State-of-the-art and prospects. *Journal of King Saud University-Computer and Information Sciences*, *34* (4), 1595–1623. https://doi.org/10.1016/j.jksuci.2020.10.012.

13. Lamaakal, I., Ouahbi, I., El Makkaoui, K., Maleh, Y., Pławiak, P., & Alblehai, F. (2024). A TinyDL model for gesture-based air handwriting Arabic numbers and simple Arabic letters recognition. *IEEE Access*, *12*, 76589–76605.

14. McEnroe, P., Wang, S., & Liyanage, M. (2022). A survey on the convergence of edge computing and AI for UAVs: Opportunities and challenges. *IEEE Internet of Things Journal*, *9* (17), 15435–15459. https://doi.org/10.1109/JIOT.2022.3176400.

15. Lamaakal, I., El Mourabit, N., El Makkaoui, K., Ouahbi, I., & Maleh, Y. (2024, October). Efficient gesture-based recognition of tifinagh characters in air handwriting with a TinyDL model. In *2024 Sixth International Conference on Intelligent Computing in Data Sciences (ICDS)* (pp. 1–8), Marrakech, Morocco. IEEE.

16. Essahraui, S., Bouyardan, M., El Hamly, I., El Makkaoui, K., Ouahbi, I., & Filali Bouami, M. (2024). Enhancing exam integrity in Moroccan higher education: An AI-based fingerprint verification model. In *Proceedings of the 7th International Conference on Networking, Intelligent Systems and Security* (pp. 1–5), Meknes, Morocco.

17. Essahraui, S., El Makkaoui, K., Filali Bouami, M., & Ouahbi, I. (2024). CNN and Vision Transformer models for detecting cheating in online examinations: A comparative evaluation. In *The International Conference on Artificial Intelligence and Smart Environment* (pp. 295–301), Errachidia, Morocco. Springer.

18. Essahraui, S., El Mrabet, M. A., Bouami, M. F., El Makkaoui, K., & Faize, A. (2022). An intelligent anti-cheating model in education exams. In *2022 5th International Conference on Advanced Communication Technologies and Networking (CommNet)*, Marrakech, Morocco, December 12–14, 2022. https://doi.org/10.1109/CommNet56067.2022.9993953.

19. Yazid, Y., Ez-Zazi, I., Guerrero-González, A., El Oualkadi, A., & Arioua, M. (2021). UAV-enabled mobile edge-computing for IoT based on AI: A comprehensive review. *Drones*, *5* (4), 148. https://doi.org/10.3390/drones5040148.

20. Essahraui, S., Ouahbi, I., El Makkaoui, K., & Filali Bouami, M. (2024). A deep learning-driven fingerprint verification model for enhancing exam integrity in Moroccan higher education. *Information Security Journal: A Global Perspective*, *33*, 1–13. https://doi.org/10.1080/19393555.2024.2374247.

21. Saha, B., Samanta, R., Roy, R. B., & Ghosh, S. K. (2025). TinyTNAS: Time-bound, GPU-independent hardware-aware neural architecture search for TinyML time series classification. *IEEE Embedded Systems Letters*. https://doi.org/10.1109/LES.2025.3561870.

3 TinyML and Edge AI for Low-Power IoT and LPWAN Applications

Ismail Lamaakal, Yassine Maleh,
Khalid El Makkaoui, and Ibrahim Ouahbi

3.1 INTRODUCTION

The Internet of Things [1] has rapidly evolved, impacting numerous sectors, including smart cities, manufacturing, healthcare, logistics, and environmental management. This expansion is driven by the integration of assets and devices through wired and wireless Internet connections, enabling data collection, automation, and intelligent control in a wide range of applications.

Given the diversity of IoT implementations, their technical requirements differ greatly [2]. They may involve various communication technologies—ranging from proximity-based and local networks to wide-area networks (WANs). In large-scale deployments, challenges such as energy efficiency, coverage, cost, and network capacity become critical concerns [3]. To address these, LPWANs have emerged, emphasizing long-range, low-energy communication with minimal bandwidth needs [4]. LPWAN technologies—such as LoRaWAN, NB-IoT, LTE-M, Sigfox, and Ingenu [5]—are increasingly adopted due to their scalability, long battery life, affordability, and, in some cases, license-free operation.

Meanwhile, progress in artificial intelligence (AI), machine learning (ML) [6], and data science has transformed how data is processed and utilized. Traditional IoT solutions rely on cloud-based processing, which introduces concerns such as high latency, limited bandwidth, and vulnerabilities in privacy and data security. Additionally, IoT devices are often constrained by limited processing power, memory, and energy capacity [7].

Edge and fog computing provide a compelling alternative by enabling local data processing closer to the source. These paradigms reduce reliance on the cloud by introducing intelligent fog nodes—routers [8,9], gateways, or clusters—capable of running AI models. This architecture minimizes network load, improves latency, and strengthens data security while offering better energy efficiency.

The push toward more compact and intelligent hardware has made it feasible to deploy ML models directly on embedded devices. These devices typically use microcontrollers (MCUs) or microprocessors with constrained computational and storage resources [10]. Although traditional ML models often exceed the limits of such devices, a new approach known as TinyML has emerged. TinyML enables the

DOI: 10.1201/9781003544449-3

deployment of optimized ML algorithms on resource-limited edge nodes, making localized intelligence feasible.

This innovation supports basic analytics at the edge with enhanced efficiency and security. As microprocessor performance and accessibility improve, the cost and complexity of deploying ML models have dropped significantly. This has spurred the growth of novel applications, including image and audio processing, event detection, and vibration analysis.

Unlike cloud-based systems, TinyML solutions reduce the need to transmit raw data, thereby decreasing latency and energy use while improving privacy. The global edge artificial intelligence (edge AI) market reflects this momentum, with projections showing a rise from $15.6 billion in 2022 to $107.47 billion by 2029—a compound annual growth rate of 31.7%. Gartner also forecasts that by 2025, over 55% of deep learning analytics will occur at the edge.[1]

3.2 IOT AND LPWAN ARCHITECTURES

Contemporary IoT systems that utilize LPWANs [11] commonly adopt star or mesh network topologies. In particular, star and star-on-star configurations enhance communication coverage and expand cell size. Within these architectures, gateway nodes operate as central hubs, relaying messages between edge devices and the cloud (see Figure 3.1). This central role places a significant energy demand on gateways, making power supply a critical factor unless they are connected to a permanent power source.

Gateways are designed to facilitate bidirectional communication with multiple end nodes. These end devices—often microcontroller-based embedded systems—are

User Monitoring Control

Cloud Application Servers
• CloudML

EdegAI/Computing Network Servers
• EdgeML (Routers, Switches, Gateways, Concentrators, Coordinators, etc.)

IoT/LPWAN Nodes
• EdgeAI: TinyML Embedded Systems (Resource-Constrained LoRa Nodes)

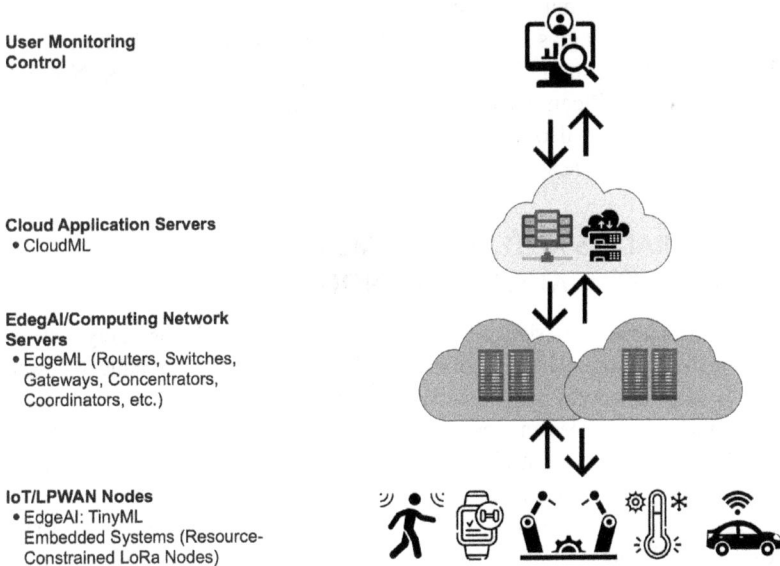

FIGURE 3.1   TinyML framework for LoRaWAN-based LPWAN.

typically located in remote areas and perform the task of sensing environmental or operational parameters. Data packets generated by these nodes are transmitted wirelessly and picked up by any reachable gateway. These gateways then forward the packets to a centralized network server located in the cloud via backhaul links, such as cellular, Wi-Fi, Ethernet, or satellite. Notably, end-to-end direct communication between nodes is not supported in such architectures.

Traditionally, this architecture leads to raw data being transmitted to the cloud for analysis, and the results are subsequently sent back to the original devices. This process introduces substantial latency and increases network traffic, both of which contribute to higher energy consumption and raise concerns over data privacy and system security.

To mitigate these limitations, lightweight ML models—developed under the TinyML paradigm—can be deployed directly on end devices. This approach enables basic data processing locally, reducing the reliance on cloud infrastructure. For instance, in LoRaWAN-based LPWANs, TinyML can be embedded into constrained IoT nodes that transmit data to multiple gateways simultaneously. As long as at least one gateway is reachable, the system can function reliably.

The gateways forward received packets to a centralized LPWAN network server, which is then connected to various application servers through the Internet. This server is responsible for packet deduplication, security validation, and integrity checking. It filters redundant data from multiple gateways and ensures secure transmission to the appropriate application server. In scenarios where packets target specific applications, the network server directs them accordingly.

One notable advantage of this architecture is the elimination of hand-off mechanisms. Since all reachable gateways can forward identical packets to the server, the system is particularly well-suited for mobile use cases, such as asset tracking, where devices frequently move across different zones.

As LPWAN technologies continue to evolve, their deployment becomes increasingly heterogeneous. A single IoT network may integrate multiple LPWAN standards, each with unique specifications and operational principles. While this enhances flexibility and scalability, it also introduces challenges in interoperability and seamless integration among diverse systems.

3.3   CONSTRAINTS AND REQUIREMENTS FOR LPWAN APPLICATIONS AND NODES

One of the most critical design constraints for LPWAN nodes is energy efficiency. Maximizing the operational lifespan of these nodes—especially when battery-powered—necessitates adopting energy-saving mechanisms. Common techniques include extremely low transmission duty cycles and restricted data rates. Moreover, LPWANs typically use lightweight medium access control (MAC) protocols, making them especially well-suited for applications with minimal data throughput requirements [11].

To further improve efficiency, minimize latency, and enhance data security at a low cost, TinyML presents itself as an effective solution. By enabling lightweight ML directly on edge devices, TinyML allows for on-node decision-making and analytics. These operations occur without needing constant communication with the cloud, thereby reducing power consumption, communication overhead, and exposure to security vulnerabilities.

TinyML is expected to play a transformative role in the emerging Internet of Everything (IoE) [12] and immersive technologies such as augmented, virtual, and extended reality (AR/VR/XR). These domains require localized, low-cost, and energy-conscious computation to support real-time, interactive experiences.

As LPWAN adoption accelerates, its use is extending into a diverse array of novel applications—many of which require stringent specifications, including low latency, low power, high data privacy, and minimal cost. Examples include asset tracking, environmental monitoring, and smart metering, as outlined in Table 3.1.

However, these use cases vary significantly in their operational demands. While some applications—such as utility metering—can standardize on a single device type, others require more heterogeneous device configurations to satisfy their complex requirements. Consequently, LPWAN solutions must be tailored to specific or broad sets of constraints, balancing factors like coverage, capacity, cost, and energy use [13].

In parallel, selecting the appropriate edge AI platform for a given application involves evaluating several parameters: computational latency, processing capacity, power efficiency, and physical characteristics such as weight and form factor. For real-time systems, platforms with high processing power and minimal delay are essential.

TABLE 3.1
Applications of IoT/LPWANs Based on TinyML

Field	Applications
Smart cities	Smart parking, structural health monitoring of buildings and bridges, air quality measurement, noise level monitoring, traffic congestion and light control, road tolling, smart lighting, trash collection, waste management, street cleaning, utility metering, fire detection, elevator monitoring, manhole cover monitoring, flood management, construction site equipment tracking, labor health monitoring, and public safety systems.
Smart environment	Monitoring water quality, air pollution, temperature changes, forest fires, landslides, animal movement, snow levels, and early earthquake detection.
Smart water	Water quality analysis, leak detection, flood monitoring, swimming pool management, and detection of chemical spills.
Smart metering	Smart meters for electricity, gas, and water; gas pipeline monitoring; and warehouse environmental monitoring.
Smart grid and energy	Grid control, load balancing, remote measurement and monitoring, transformer health analysis, and monitoring wind/solar installations.
Security and emergencies	Perimeter access control, liquid detection, radiation monitoring, and detection of explosive or hazardous gases.
Retail	Supply chain optimization, smart shopping systems, intelligent shelves, and automated product management.

(Continued)

TABLE 3.1 (*Continued*)
Applications of IoT/LPWANs Based on TinyML

Field	Applications
Transport and logistics	Insurance tracking, vehicle monitoring, rental/lease systems, road hazard detection, localization, routing, shipment condition monitoring, storage compatibility checks, fleet tracking, smart trains, and mobility-as-a-service systems.
Industrial automation and smart manufacturing	Machine-to-machine systems, robotics, indoor air and temperature monitoring, production line control, ozone monitoring, indoor localization, vehicle diagnostics, predictive maintenance, energy usage tracking, and industrial-as-a-service platforms.
Smart agriculture and farming	Monitoring temperature, humidity, alkalinity; wine quality optimization; smart greenhouses; robotic agriculture; weather stations; compost/fertilizer control; hydroponics; livestock tracking; and gas toxicity monitoring.
Smart homes/buildings and real estate	Monitoring energy/water use, temperature, humidity, smoke/CO/CO_2 levels, floods, occupancy detection (space-as-a-service), intrusion detection, remote control of appliances, and preservation of art and goods.
eHealth, life sciences, and wearables	Monitoring patient health, connected medical environments, wearable sensors, patient surveillance, UV exposure monitoring, telemedicine, fall detection, assisted living, medical storage conditions, athletic health tracking, chronic disease monitoring, and mosquito/insect population tracking.
Agriculture and food	Fish farming optimization, animal welfare, farm air quality monitoring, irrigation water monitoring, solar radiation detection, yield maximization, and pest/disease prevention.
Sustainability	Air and water quality monitoring in natural areas, forest fire detection, and flood risk assessment.

Both LPWAN and edge AI infrastructures can be fine-tuned based on the application context. Table 3.2 presents a comparative mapping of various use cases with their respective priorities for LPWAN and edge AI. It assigns qualitative values—high (H), medium (M), or low (L)—to key design considerations such as interoperability, latency, and energy demands. This decision matrix helps architects and developers select optimal configurations tailored to their specific deployment goals.

3.4 EDGE AI AND TinyML

Recent advancements in AI and ML have led to profound transformations across numerous sectors, including industry, business, healthcare, and daily life. These breakthroughs have unlocked new opportunities while simultaneously introducing technical and ethical challenges.

Traditionally, cloud-based AI has served as the primary platform for executing complex ML models. In this paradigm, raw data collected by IoT or edge devices is transmitted to remote servers where AI models process the data. Once analyzed,

TABLE 3.2

Mapping Applications with Their Requirements for LPWAN and Edge AI

Applications Headergray	LPWAN Requirements						Edge AI Requirements		
	Coverage	Capacity	Cost	Low Power	Specific Needs	Latency	Processing Power	Low power	Size/weigh
Smart cities	H	H	H	M	H	H	M	M	H
Smart environment Smart water	M H H H H	H M H H L	H M H M M	H M M M H	M L L M H	M M M H H	M L L H H	H M M M H	M M M M H
Smart metering	H H	H H	H M	L L	M H	M H	M H	L L	H H
Smart grid and energy Security and emergencies	L	H	H	L	L	H	H	L	M
Retail automotives and logistics									
Industrial automation and smart manufacturing	H H H H	H M H H	M L M H	H L H H	L L H H	M H H H	M H H M	H L H H	H M H H
Smart agriculture and farming, food									
Smart homes/building and real estate eHealth, life sciences, and wearables									
Sustainable applications									

the results are sent back to the original devices for further action. While effective for high-complexity tasks, this method suffers from several limitations: increased latency, significant bandwidth requirements, elevated power consumption, and heightened concerns regarding data security and privacy.

This model becomes even less practical given the growing prevalence of resource-constrained IoT and LPWAN nodes. These devices—often powered by microcontrollers—possess limited computational capabilities, minimal memory, and are typically battery-operated. Their primary role is to capture sensor data and perform basic processing. Reliance on cloud processing can introduce unacceptable delays and deplete power resources rapidly.

To overcome these challenges, the concept of edge AI has emerged [14]. This approach enables AI model execution directly on the device, thereby eliminating the need for frequent cloud communication. Edge AI facilitates real-time decision-making by processing data locally, reducing latency, improving data privacy and security, lowering power consumption, and decreasing reliance on network connectivity.

A specialized subfield of edge AI, known as TinyML, has gained traction for ultra-low-power applications. TinyML focuses on deploying highly optimized ML models on microcontroller-based systems. This paradigm enables on-device analytics at an extremely low cost and energy consumption, without compromising core functionality.

The benefits of edge AI and TinyML are numerous: low latency, reduced bandwidth demand, enhanced privacy and data protection, minimal energy usage, local learning capabilities, and overall cost efficiency [15]. These advantages make them ideal for the expanding IoE landscape, where billions of intelligent but constrained devices operate autonomously.

As technological trends continue to evolve, many AI applications are expected to shift toward the edge. This transition will further decentralize computation, bringing intelligence closer to the data source. Despite their promise, edge AI and TinyML also present several technical challenges—particularly due to hardware limitations and system integration constraints—which are explored in the following section.

3.5   CHALLENGES FOR EDGE AI AND TinyML

Edge AI has transformed the computing paradigm by enabling localized data processing on devices, thereby reducing dependence on centralized cloud or fog servers. Despite its benefits, the deployment of edge AI—especially in conjunction with TinyML—faces several technical and operational challenges that impact its efficiency, scalability, and reliability [16].

- **Energy-Efficient Operations:** Ensuring energy efficiency is essential for prolonging node lifespan, especially in LPWAN scenarios with infrequent, low-rate data transmission. Nodes must support ultra-low-power modes such as sleep states and deactivate high-energy components (like transceivers) during idle periods [14]. These transceivers should only be triggered for active communication. Additionally, the ML models running on these devices must be lightweight and optimized to operate within the limited energy budgets.

- **Scalability:** As the number of edge devices grows, scalability becomes a significant concern. Managing increased data volumes and network traffic can lead to reduced reliability and performance. Effective scalability demands load balancing, distributed processing strategies [17], and techniques like edge orchestration and edge-to-cloud coordination. These help maintain system responsiveness and robustness under scale.
- **Limited Memory:** Most edge devices are constrained by minimal onboard memory and limited storage capacity. As datasets grow or as more complex ML models are required, devices may fail to maintain real-time performance [17]. Therefore, choosing or designing memory-efficient algorithms and developing benchmarking datasets tailored to edge constraints are vital.
- **Latency:** Even with local processing, delays can arise from sensor input to actionable output due to factors such as memory access time, sensor response, and ML model complexity. Strategies like edge caching—storing frequently used data near the node—and federated learning—training models across distributed nodes without sharing raw data—are effective in mitigating latency while preserving privacy.
- **Machine Learning Model Complexity:** Deploying advanced ML models on embedded systems is often limited by the absence of dedicated ML accelerators. Furthermore, high-level model development tools may not directly translate to edge-compatible versions, complicating compression, portability, and execution efficiency.
- **Cost Effectiveness:** The rapid expansion of IoT and LPWANs introduces cost challenges. Edge devices often require specialized microcontrollers, sensors, and actuators, which can drive up hardware costs. On the software side, development and maintenance demand niche skills and must operate within constrained resources. Connectivity, storage, and energy infrastructure also add to the financial burden. Strategies such as leveraging existing infrastructure, adopting open-source solutions, and optimizing hardware-software integration are essential to maintain affordability.
- **Network and Data Integration:** Data management and system interoperability are major hurdles in edge environments. Variability in hardware, software, and communication protocols across devices can cause integration conflicts. Data movement at the edge is tightly linked with latency and power consumption, particularly for high-volume or high-frequency sources. Techniques such as federated learning and data compression are essential to maintain performance and privacy without overloading the network.
- **Data Security and Privacy:** Security is paramount in edge AI implementations. Devices should employ secure boot mechanisms and hardware root-of-trust to protect system integrity. These are complemented by secure software practices like threat modeling, code reviews, and encrypted communication protocols. A major advantage of TinyML is that it keeps raw data localized, thereby reducing privacy risks and exposure to cyber threats.

3.6   HARDWARE AND SOFTWARE CONSIDERATIONS

Deploying AI at the edge requires careful evaluation of both hardware and software components to meet the constraints and goals of various IoT and LPWAN applications. Numerous edge AI platforms are available to support this deployment by enabling algorithm execution closer to the data source. This proximity reduces latency, lowers bandwidth consumption, and enhances real-time responsiveness.

Popular platforms for edge AI include TensorFlow Lite, PyTorch Mobile, Open VINO, NVIDIA Jetson, Edge Impulse, Caffe2, and Apache MXNet [1]. Each of these offers different strengths in terms of model optimization, device support, and processing capabilities [2]. For TinyML specifically, TensorFlow Lite and TensorFlow Lite Micro are among the most widely adopted due to their lightweight runtime and compatibility with microcontrollers.

Hardware Considerations: Edge devices—particularly those based on microcontrollers—typically have significant limitations in terms of computational power, memory, and battery life. These limitations can impact their ability to perform complex tasks such as image processing or natural language understanding. To address these constraints:

- Specialized hardware such as field-programmable gate arrays (FPGAs), application-specific integrated circuits (ASICs), and GPUs can be used to boost processing efficiency at the edge.
- Real-time processing depends heavily on factors such as CPU clock speed, available RAM, and memory access speeds.
- Efficient power management is critical. Strategies such as low-power chipsets, hardware accelerators, dynamic frequency scaling, and optimized duty cycles help extend device lifetime without frequent recharging.

Software Considerations: Selecting appropriate software tools is equally important for the success of edge AI systems. Software must be compatible with a wide range of components, including sensors, processors, operating systems, and programming environments [3].

- Integration across hardware and software must be seamless to support real-time processing.
- Software should be scalable, capable of handling increasing data volumes and computational demands without sacrificing performance.
- Precision and reliability are essential for generating actionable insights, especially in critical applications like healthcare and smart infrastructure.
- Interpretability must also be prioritized; models should output results that are logically structured and understandable by end-users or other automated systems.

Overall, optimizing both hardware and software resources is crucial to meet the stringent requirements of edge AI and TinyML deployments. Careful co-design of these components ensures that systems remain responsive, efficient, and adaptable to the growing demands of intelligent IoT ecosystems.

3.7 CONCLUSION AND FUTURE WORK

This chapter has highlighted the transformative potential of integrating Tiny Machine Learning (TinyML) and edge AI into low-power IoT and LPWAN ecosystems. By enabling localized, energy-efficient, and real-time data processing, these technologies address key challenges in latency, bandwidth constraints, and data privacy. We examined the architectural frameworks, application domains, and technical limitations that define the current landscape, as well as the hardware and software considerations crucial for successful deployment. Despite significant progress, future research should focus on improving model compression techniques, ensuring secure and privacy-preserving learning at the edge, and developing standardized benchmarks to evaluate performance across heterogeneous platforms. Additionally, further exploration into interoperable software stacks and scalable deployment strategies will be essential to support the next generation of intelligent, resource-constrained IoT systems.

NOTE

1 https://www.gartner.com/en/newsroom/press-releases/2023-08-01-gartner-identifiestop-trends-shaping-future-of-data-science-and-machine-learning.

REFERENCES

1. I. Lamaakal et al. 'A Comprehensive Survey on Tiny Machine Learning for Human Behavior Analysis'. *IEEE Internet of Things Journal* 12 (2025), pp. 32419–32443.
2. I. Lamaakal et al. 'A TinyDL model for gesture-based air handwriting Arabic numbers and simple Arabic letters recognition'. *IEEE Access* 12 (2024), pp. 76589–76605.
3. I. Lamaakal et al. 'A TinyML model for gesture-based air handwriting Arabic numbers recognition'. *Procedia Computer Science* 236 (2024), pp. 589–596.
4. N. S. Chilamkurthy et al. 'Low-power wide-area networks: A broad overview of its different aspects'. *IEEE Access* 10 (2022), pp. 81926–81959.
5. S. Ugwuanyi, G. Paul & J. Irvine. 'Survey of IoT for developing countries: Performance analysis of LoRaWAN and cellular NB-IoT networks'. *Electronics* 10.18 (2021), p. 2224.
6. I. Lamaakal et al. 'Optimizing Breast Calcification Detection in Mammography Using PySpark: A Big Data and Machine Learning Approach'. In: A. Swaroop, B. Virdee, S. D. Correia & Z. Polkowski (eds.), *Proceedings of Data Analytics and Management. International Conference on Data Analytics & Management.* Singapore: Springer, 2025, pp. 617–627.
7. A. Al-Fuqaha et al. 'Internet of Things: A survey on enabling technologies, protocols, and applications'. *IEEE Communications Surveys & Tutorials* 17.4 (2015). Fourth Quarter, pp. 2347–2376.
8. I. Lamaakal et al. 'A deep learning-powered TinyML model for gesture-based air hand-writing simple arabic letters recognition'. In: *International Conference on Digital Technologies and Applications.* Cham: Springer Nature Switzerland, 2024, Benguerir, Morocco, pp. 32–42.
9. I. Lamaakal et al. 'Tiny language models for automation and control: Overview, potential applications, and future research directions'. *Sensors* 25.5 (2025), p. 1318.
10. I. Lamaakal et al. 'Efficient Gesture-Based Recognition of Tifinagh Characters in Air Handwriting with a TinyDL Model'. In: *2024 Sixth International Conference on Intelligent Computing in Data Sciences (ICDS)*, Marrakech, Morocco, October 23–24, 2024. IEEE, 2024, pp. 1–8.

11. B. S. Chaudhari, M. Zennaro & S. Borkar. 'LPWAN technologies: Emerging application characteristics, requirements, and design considerations'. *Future Internet* 12.3 (2020), p. 46.

12. B. S. Chaudhari et al., eds. *TinyML for Edge Intelligence in IoT and LPWAN Networks.* Elsevier, 2024.

13. R. K. Singh et al. 'Energy consumption analysis of LPWAN technologies and lifetime estimation for IoT application'. *Sensors* 20.17 (2020), p. 4794.

14. J. Sanchez-Gomez et al. 'Integrating LPWAN technologies in the 5G ecosystem: A survey on security challenges and solutions'. *IEEE Access* 8 (2020), pp. 216437–216460.

15. L. Dutta & S. Bharali. 'TinyML meets IoT: A comprehensive survey'. *Internet of Things* 16 (2021), p. 100461.

16. R. Kallimani et al. 'TinyML: Tools, applications, challenges, and future research directions'. *Multimedia Tools and Applications* 83.10 (2024), pp. 29015–29045.

17. V. Rajapakse, I. Karunanayake & N. Ahmed. 'Intelligence at the extreme edge: A survey on reformable TinyML'. *ACM Computing Surveys* 55.13s (2023), pp. 1–30.

4 Efficient Real-Time Mask Detection Using TinyML

Ismail Lamaakal, Yassine Maleh,
Khalid El Makkaoui, and Ibrahim Ouahbi

4.1 INTRODUCTION

The global outbreak of COVID-19 has brought unprecedented challenges to public health systems around the world [1–5], emphasizing the critical need for effective preventive measures. Among various interventions, mask wearing has been widely recommended by health organizations as a primary deterrent against the spread of the virus [6–9]. While the concept is straightforward, ensuring widespread compliance has proven to be a complex endeavor in public health. The integration of technology into public health initiatives provides a promising avenue to monitor and enforce these preventive measures more effectively and in real-time. Within the proposed technological solutions, TinyML [10–14] emerges as a particularly valuable tool due to its suitability for deployment in resource-constrained environments.

TinyML [15] allows for the development of smart, embedded systems capable of performing data processing tasks directly on MCUs [16] without the need for continuous cloud connectivity. This is crucial in public settings where constant data transmission is impractical or privacy concerns are significant. The application of TinyML [17] in public health [18,19], particularly for real-time detection tasks like mask adherence, requires models that are efficient in both computational resource usage and power consumption. This efficiency must be achieved without sacrificing the accuracy or responsiveness of the system, as public health decisions depend on reliable and timely data.

Despite significant advancements in ML and its applications in healthcare [20–22], the use of TinyML for real-time public health monitoring under pandemic conditions is relatively unexplored [23–27]. Existing models often struggle with the dual challenge of operating within the low-power, low-memory constraints of MCUs while requiring minimal supervised training data to achieve high accuracy. There is a significant gap in the development of TinyML models that can be effectively implemented in everyday environments, specifically for tasks as critical as mask detection during a pandemic.

This chapter introduces a cutting-edge solution that leverages Vision Wake Words (VWW) technology combined with the MobileNet architecture [28], known for its depthwise separable convolutions, to create a highly efficient and accurate mask detection system. By utilizing transfer learning, we have significantly reduced the model's dependency on large training datasets, thereby decreasing the training time

DOI: 10.1201/9781003544449-4

while maintaining high performance. Our approach successfully compresses the original 8.62 MB model to a mere 214 KB using TensorFlow Lite with Quantization, making it feasible for deployment on the most constrained MCUs. The final model demonstrates a 99.6% accuracy rate in detecting mask usage, offering a robust tool for enhancing public health safety. These advancements contribute significantly to Healthcare 4.0, integrating smart sensors and IoT to enhance healthcare service delivery and compliance with health guidelines.

This chapter presents several significant advancements in the application of machine learning to public health crises. The key contributions include the following:

1. **Innovative Application of TinyML:** Introduction of a novel application of TinyML for real-time mask detection, integrating advanced machine learning algorithms with ultra-low-power embedded systems for scalable public health solutions.
2. **Enhancing Public Health Compliance:** Presentation of a technology-driven approach to enforce mask compliance efficiently during health crises, establishing a framework for real-time, non-invasive monitoring of public health measures to increase compliance potentially.
3. **Resource-Efficient Technology for Health Surveillance:** Addressing the challenge of resource scarcity by integrating the MobileNet architecture with TinyML, which provides a pathway for deploying advanced health surveillance technologies in settings where computational resources are limited.
4. **High Accuracy in Mask Detection with Low Resource Use:** Detailing the development of a high-accuracy, TinyML-based mask detection system that operates efficiently on low-power devices, suitable for widespread implementation in critical infrastructures and public spaces.

The structure of our chapter is designed to provide a comprehensive overview of our research and findings. In Section 4.2, we delve into the relevant literature, presenting a review of related work that sets the context for our novel contributions. In Section 4.3, we present an in-depth examination of the MobileNetV2 architecture, focusing on its efficiency for mask detection. We discuss the architectural details of MobileNetV2 and highlight the modifications made to optimize its performance for our specific task. This is followed by a detailed description of our methodology in Section 4.4, where we outline the processes involved in preparing our dataset, data preprocessing techniques, modifications made to the MobileNetV2 architecture, the methods used for model conversion to TensorFlow Lite, and the strategies for deploying the model on constrained devices. Section 4.5 is devoted to a rigorous analysis of our findings and results. Here, we discuss the strengths and limitations of our approach, providing insights into the accuracy and efficiency of the mask detection model. We also explore the practical implications of our research in Section 4.6, discussing the feasibility of implementing our model in real-world scenarios with considerations of cost-effectiveness. The chapter concludes with Section 4.7, where we summarize our key conclusions and discuss potential avenues for future work, aiming to enhance further the scalability and impact of TinyML in public health initiatives.

4.2 RELATED WORK

Deep learning models have been shown to outperform shallow models in complex object identification tasks [29]. One example is developing a real-time system or model capable of detecting whether or not people have used masks in public places.

The authors of Ref. [30] employed real-time deep learning to classify and distinguish emotions, whereas VGG-16 was used to categorize seven faces. This method was particularly effective during the Covid-19 lock-down period to limit the spread of cases. Ejaz et al. [31] employed principal component analysis to distinguish between people with masked and unmasked faces.

The authors of Ref. [32] used a CNN to track and enforce compliance with health guidelines by identifying mask wearers. The HGL approach classifies head postures using masks for faces by analyzing color texture in pictures and line portraits. Front accuracy was 93.64%, with side-to-side accuracy at 87.17%.

The authors of Ref. [33] developed a face mask recognition design utilizing the condition identification approach. The research divided the task into four steps: pre-processing the image, clipping face areas, super-resolution operation, and estimating the final state. The study's main novelty was the use of super-resolution to enhance low-quality picture performance. The author's suggested technique, which utilized SRCNet, achieved 98.7% accuracy in detecting face masks and their positions.

The authors of Ref. [34] developed a GAN-based system for detecting face masks, synthesizing missing facial components, and reconstructing areas. The suggested GAN used two discriminators: the first analyzed the structure of the face mask, and the second extracted the region hidden by the mask. They trained the model using two fake datasets.

In Ref. [35], the authors use the Darknet-53 (YOLOv3) algorithm for facial detection.

The authors of Ref. [36] proposed a mobile phone-based detection technique. Three components were derived from GLCMs of face mask micro-images. A three-result recognition analysis utilizing the KNN algorithm showed an overall accuracy of 82.87%. To distinguish face masks in microphotos, the system employed a gray-level co-occurrence matrix. However, the approach is confined to smartphones and may not be appropriate for all applications.

The authors of Ref. [37] recommended employing a pretrained MobileNet with the global pooling block for facial recognition and detection. The pre-configured MobileNet creates a multidimensional component map from a shaded image. The suggested model's usage of an overall pooling block prevents overfitting. The authors of Ref. [38] demonstrate an innovative hybrid model combining deep transfer learning with machine learning to detect face masks during the COVID-19 pandemic efficiently. This approach enhances accuracy and adaptability in real-world applications, offering significant improvements over traditional deep learning methods.

The authors of Ref. [39] present a novel approach to enhance facial recognition technology under masked conditions, a pertinent topic due to the COVID-19 pandemic. Utilizing advanced deep learning algorithms, this chapter explores the integration of innovative neural network architectures to improve the robustness and accuracy of mask detection systems.

The study by Ref. [40] introduces the "Self-restrained Triplet Loss" method to enhance the accuracy of masked face recognition, an essential adaptation due to the widespread use of face masks during the COVID-19 pandemic. This method innovatively modifies the conventional triplet loss approach, optimizing it to better handle the challenges posed by partially obscured facial features. The authors of Ref. [41] tackle the obstacles in masked face recognition (MFR) due to mask occlusion by proposing a new method, masked face data uncertainty learning (MaskDUF). This method adjusts optimization weights based on sample uncertainty and recognizability, achieving better class distinction and robustness. Their approach demonstrates notable improvements in accuracy compared to existing models.

The study [42] introduces a lightweight CNN model to tackle the challenge of recognizing masked faces during the COVID-19 pandemic, using the HSTU Masked Face Dataset. Enhanced with batch normalization, dropout, and depthwise normalization, this model surpasses traditional models like VGG16 and MobileNet with a 97% recognition accuracy.

4.3 MOBILENETV2 ARCHITECTURE: AN EFFICIENCY OVERVIEW

The MobileNetV2 network, which has been pre-loaded using the ImageNet database [43], serves as the basis for this modified model. It is capable of categorizing images into various object classes. In this adaptation, MobileNetV2 is implemented as a transfer learning model, while a Caffe model is employed for single-shot detection (SSD) tasks like identification and verification. MobileNetV2, an advancement over its predecessor MobileNetV1, incorporates inverted residual blocks with linear bottlenecks and boasts a reduced number of parameters, enhancing its efficiency. The architecture supports input sizes larger than 32×32, and performance improves with larger image dimensions. Figure 4.1 depicts the architectural schematic of MobileNetV2, resembling a CNN-based deep learning framework that includes layers for convolution, pooling, dropout, non-linear functions, full connectivity, and

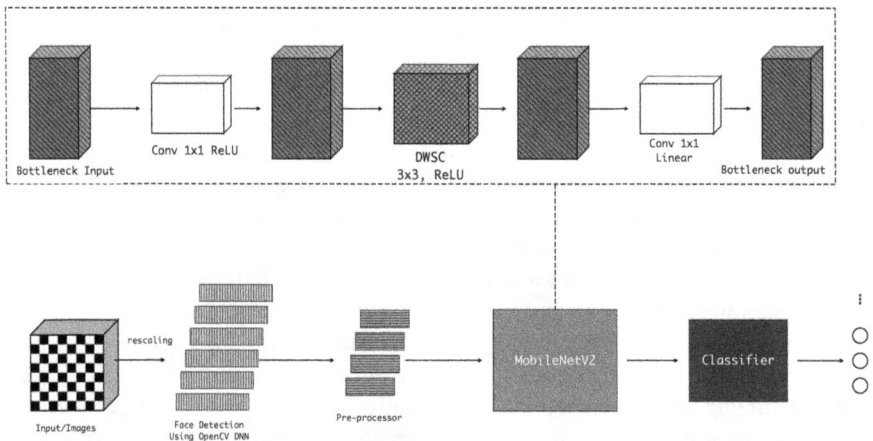

FIGURE 4.1 MobileNetV2 architecture.

linear bottlenecks [44]. The structure features a 1×1 convolution layer, seventeen 3×3 convolutional layers, a maximum pooling average layer, and a classification layer. For effective training and accurate mask detection, it is crucial to utilize substantial datasets. The primary objective is to enhance the accuracy of face mask detection while minimizing the number of trainable parameters. A key component of MobileNetV2 is the depthwise separable convolution (DWSC) layer, which significantly speeds up the process. MobileNets significantly minimize the scale of neural networks used in computer vision. Let's explore further into their main innovative feature: depthwise separable convolutions.

To grasp the distinctions between depthwise separable convolutions and standard convolutions, we should first revisit and analyze standard convolutions. Specifically, we'll focus on calculating the number of multiplication operations and parameters involved in a standard convolutional operation.

Typically, our focus has been on filters used on grayscale images, such as the vertical line filter, as illustrated in Figure 4.2.

These images are also known as single-channel images because they contain only one set of pixel values. In contrast, most color images are composed of three channels—RGB—each representing the intensity levels of red, green, and blue. Consequently, the filters (often called kernels) used are more complex than a mere matrix; they are tensors that must operate across all three channels simultaneously. Below, we demonstrate this tensor operation. In the provided example, a $3 \times 3 \times 3$ kernel convolves with a $9 \times 9 \times 3$ image, resulting in a $7 \times 7 \times 1$ output. The highlighted squares indicate the inputs and outputs from the final convolution operation (see Figure 4.3).

Typically, the convolution process necessitates $D_K \times D_K \times M$ multiplicative operations, considering the kernel's dimensions (D_K) and the count of image channels (M). To generate a singular output point, these calculations are reiterated $D_F \times D_F$ times over the output feature map. In standard practice, a set of N kernels, collectively

FIGURE 4.2 Application of a vertical line filter on a grayscale image.

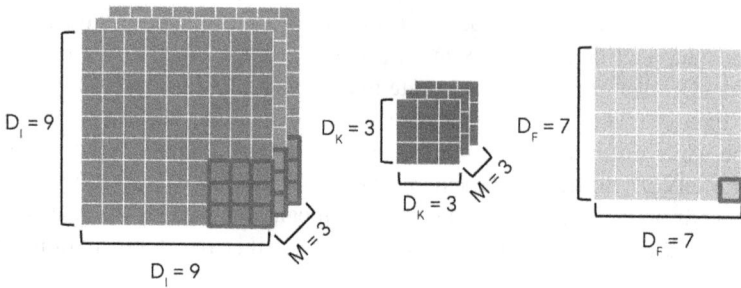

FIGURE 4.3 Convolution operation: 3D kernel applied to an RGB image.

termed as a filter, is utilized instead of a solitary kernel. A filter comprises various kernels concatenated together, with each kernel tailored to process a specific channel of the input data. Hence, the aggregate of multiplication operations required is encapsulated by the formula:

$$D_K * D_K * M * D_F * D_F * N = D^2 * M * D^2 * N. \tag{4.1}$$

4.3.1 Depthwise Separable Convolutions

Depthwise separable convolutions unfold through a two-stage procedure. Initially, individual channels are processed independently, akin to solitary grayscale images, with the application of respective filters generating multiple intermediary outputs—this stage is termed as depthwise convolution. Subsequently, a pointwise convolution is executed on these intermediary outputs, utilizing a $1 \times 1 \times C$ filter to derive the conclusive output. Illustrated below is an example of this method where a $9 \times 9 \times 3$ image undergoes convolution with three distinct 3×3 filters on a depthwise basis, yielding a $7 \times 7 \times 3$ intermediary. Following this, a $1 \times 1 \times 3$ pointwise filter is employed to generate the final $7 \times 7 \times 1$ output as shown in Figure 4.4.

Each depthwise convolution demands M filters carrying out $D_K \times D_K$ multiplications. For generating a single output feature map, these operations are conducted $D_F \times D_F$ times. This results in a cumulative count of multiplications for the depthwise phase being $D_K \times D_K \times M \times D_F \times D_F$.

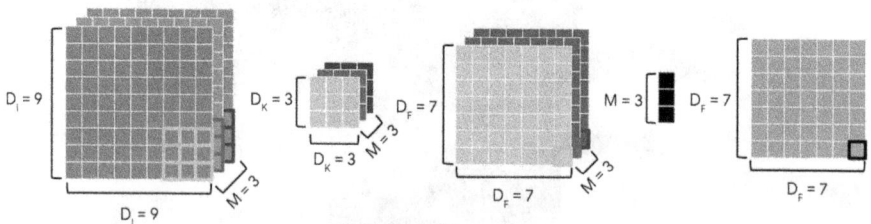

FIGURE 4.4 Depthwise and pointwise convolutional process on a $9 \times 9 \times 3$ image.

For pointwise convolutions, we apply a $1\times1\times M$ filter over the feature map's depth $D_F\times D_F$ times. Contrary to conventional convolutions that utilize just one filter, here multiple filters are sequentially applied. In the scenario of depthwise separable convolutions, these numerous filters are utilized during the pointwise phase. Thus, if there are N pointwise filters, then $M\times D_F\times D_F\times N$ multiplications are required at this stage.

Adding up the multiplications needed for both stages, the overall required number of operations is computed as

$$D_K\times D_K\times M\times D_F\times D_F + M\times D_F\times D_F\times N = M\times D_F^2\times\left(D_K^2+N\right). \quad (4.2)$$

Comparing the two types of convolutional computations, we determine their efficiency by analyzing the ratio of the required multiplications for each. By assigning standard convolutions to the denominator of our ratio, we derive

$$\frac{\text{Depthwise Separable}}{\text{Standard}} = \frac{M\times D_F^2\times\left(D_K^2+N\right)}{D_K^2\times M\times D_F^2\times N}. \quad (4.3)$$

This can be simplified to

$$\frac{\text{Depthwise Separable}}{\text{Standard}} = \frac{D_K^2+N}{D_K^2\times N}. \quad (4.4)$$

The standard convolution operation can be succinctly represented as:

$$\frac{\text{Depthwise Separable}}{\text{Standard}} = \frac{1}{N}+\frac{1}{D_K^2}. \quad (4.5)$$

This infers that utilizing more filters and larger kernels equates to a reduction in multiplication operations. For instance, with $D_K=3$ and a conservative estimate of $N=10$ filters, the ratio becomes approximately 0.2111, illustrating that depthwise separable convolutions result in a saving of nearly fivefold in multiplication operations, thus markedly improving efficiency and potentially lessening latency.

Observe that in standard convolution operations, we possess $D_K^2\times M\times N_K$ learnable parameters within our assortment of filters/kernels. In contrast, depthwise separable convolution has a parameter quantity of $D_K^2\times M + M\times N$ considering the ratio of the two types:

$$\frac{\text{Depthwise Separable}}{\text{Standard}} = \frac{1}{N}+\frac{1}{D_K^2}. \quad (4.6)$$

This highlights that depthwise separable convolution necessitates considerably less memory, as it has far fewer parameters to store.

Nonetheless, this efficiency comes with a trade-off. By curtailing the number of parameters to enhance latency and memory requirements, our models inherently sacrifice some degree of expressiveness. This limitation is usually acceptable for TinyML applications but should be taken into account when employing depthwise separable convolutions more broadly.

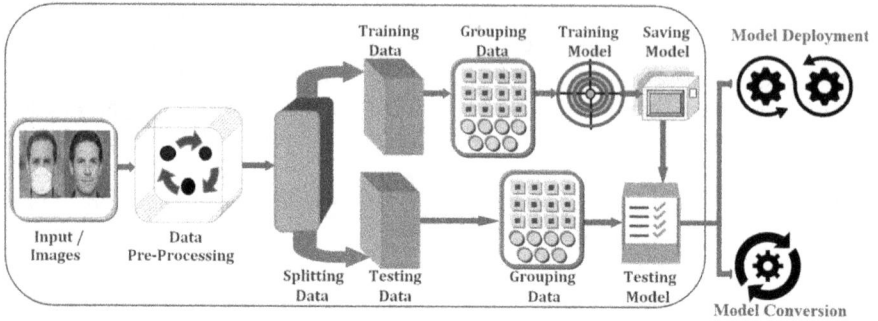

FIGURE 4.5   Schematic overview of methodology.

4.4 METHODOLOGY

In this section of our research, we outline the systematic approach employed to engineer a TinyML model, designed to facilitate facial recognition tasks, including the critical ability to discern whether individuals are wearing masks. Figure 4.5 illustrates the transition from initial data acquisition through to the ultimate deployment of the model, ensuring a detailed walkthrough of data preprocessing, division into training and testing datasets, model training with rigorous validation, model conversion, and subsequent deployment. This section is crafted to impart a transparent and detailed framework of our experimental strategy, underscoring the procedural nuances that fortify the reliability and performance of our TinyML-based facial recognition system, tailored for low-resource environments inherent to our study's focus.

4.4.1 DATA DESCRIPTION

In our research, we have utilized the Face Mask Lite Dataset [45] created by Prasoon Kottarathil. This dataset has been instrumental in advancing our studies on face mask detection technologies, providing a robust foundation of 20,000 high-definition images, evenly split between individuals wearing masks and those without. Each image in this dataset was generated using the sophisticated StyleGAN-2 [46], ensuring the high quality and consistency needed for effective training and evaluation of machine learning models in public health applications. Figure 4.6 presents a selection of examples from the dataset, showcasing images of individuals with and without masks.

4.4.2 DATA PREPROCESSING

In this section of our research, we focus on the preprocessing phase, a crucial step to prepare the dataset for the subsequent machine learning tasks. This phase encompasses several key procedures, including data augmentation, image resizing, normalization, and ultimately, data splitting into training and test sets.

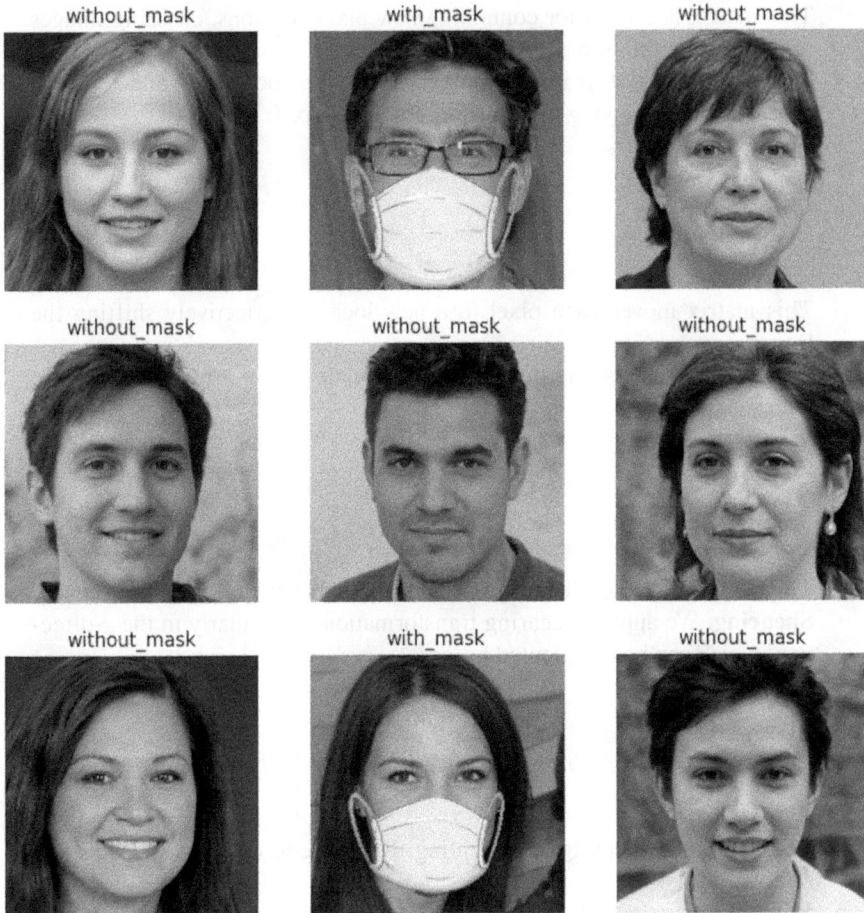

FIGURE 4.6   Dataset samples: masked and unmasked images.

4.4.2.1   Data Augmentation

In our research, we employ data augmentation [47] to enhance the diversity of our image dataset, applying various transformations to train our model under different conditions. Here's a detailed explanation of common data augmentation techniques we use

- **Rotation**: We rotate the images by a certain angle θ, using the rotation transformation matrix M:

$$M = \begin{bmatrix} \cos\theta & -\sin\theta & 0 \\ \sin\theta & \cos\theta & 0 \\ 0 & 0 & 1 \end{bmatrix}.$$

This matrix is crucial for computing new pixel positions, ensuring images remain within the frame post-rotation.

- **Translation**: We shift images in the x and y directions by t_x and t_y pixels, respectively, employing the transformation matrix T:

$$T = \begin{bmatrix} 1 & 0 & t_x \\ 0 & 1 & t_y \\ 0 & 0 & 1 \end{bmatrix}.$$

This matrix moves each pixel to a new location, effectively shifting the image.

- **Scaling**: We change the image size by a scale factor s, using the scaling matrix S:

$$S = \begin{bmatrix} s & 0 & 0 \\ 0 & s & 0 \\ 0 & 0 & 1 \end{bmatrix}.$$

This uniformly scales the image dimensions, either enlarging or shrinking it.

- **Shearing**: We apply a shearing transformation, particularly in the x-direction, which can be represented by

$$Sh_x = \begin{bmatrix} 1 & \tan(\phi) & 0 \\ 0 & 1 & 0 \\ 0 & 0 & 1 \end{bmatrix}.$$

where ϕ is the shearing angle, tilting the image to create a "windblown" effect.

- **Flip**: Flipping images horizontally or vertically involves mirroring the image across its vertical or horizontal axis, respectively.
- **Color Augmentation**: We adjust the image colors, including brightness, contrast, saturation, and hue alterations. These are calculated by adjusting pixel values directly.
- **Normalization**: We normalize pixel values to a mean of zero and a standard deviation of one:

$$\text{Normalized Pixel Value} = \frac{\text{Pixel Value} - \mu}{\sigma}, \tag{4.7}$$

where μ is the mean and σ is the standard deviation of pixel values.

By integrating these transformations randomly across different images in our dataset, we substantially improve our model's ability to generalize, which is crucial for robust performance on new, unseen data.

Following the data augmentation stage, our dataset expanded to include a total of 46,233 images.

4.4.2.2 Image Resizing

In this research endeavor, our focus extended to the meticulous resizing of images, ensuring their dimensions spanned across four distinct resolutions: 128×128, 160×160, 192×192, and 224×224. This deliberate variation was not arbitrary; rather, it served as a deliberate exploration into the impact of image resizing on the efficacy of our TinyML model. By subjecting the resized images to rigorous analysis, we aimed to uncover nuanced insights into how alterations in resolution affect model performance across a spectrum of metrics. This comprehensive examination is poised to enrich our understanding of the interplay between image dimensions and model outcomes, thereby informing future advancements in TinyML deployment within the realm of image recognition.

4.4.2.3 Data Splitting

In this study, our approach to data splitting involves a meticulous partitioning of the dataset into three distinct subsets. We allocate 80% of the dataset for training, 10% for validation, and the remaining 10% for testing purposes. This strategic division ensures that our model is trained on a substantial portion of the data while reserving separate subsets for validation and testing. The training set serves as the foundation for model learning, enabling it to capture underlying patterns and relationships within the data. The validation set plays a critical role in fine-tuning model parameters and evaluating its performance during the training process. Finally, the testing set offers an unbiased evaluation of the model's generalization capabilities on unseen data, thereby ensuring robustness and reliability in real-world scenarios. Through this meticulous data-splitting approach, we aim to develop a well-validated and dependable TinyML model for accurate image recognition tasks.

4.4.2.4 The Proposed TinyML Model

In developing our TinyML model for mask detection, we leverage the MobileNetV2 architecture as a feature extractor coupled with transfer learning techniques. This approach enables us to leverage a pretrained MobileNetV2 network and tailor it to our specific task of classifying individuals as wearing masks or not.

4.4.2.4.1 Create the Base Model

We will construct the foundational model using Google's MobileNetV2, which has been pretrained on the vast ImageNet dataset containing 1.4 million images across 1,000 classes. This extensive dataset, covering a broad spectrum of categories from jackfruit to syringe, provides a rich base for understanding and categorizing images. Such a diverse foundation will assist us in the specific task of discerning whether individuals in our dataset are wearing masks.

Initially, we must decide which layer of MobileNetV2 will serve as the repository of high-level features for our purpose. Our goal is to repurpose the model's output for a new task; hence, we focus on the final layer before the classification, the layer often referred to as the "bottleneck layer" in some literature. This layer is typically found just before the flattening process in many image processing models. Conceptually, in machine learning models, inputs are perceived at the bottom while outputs are at

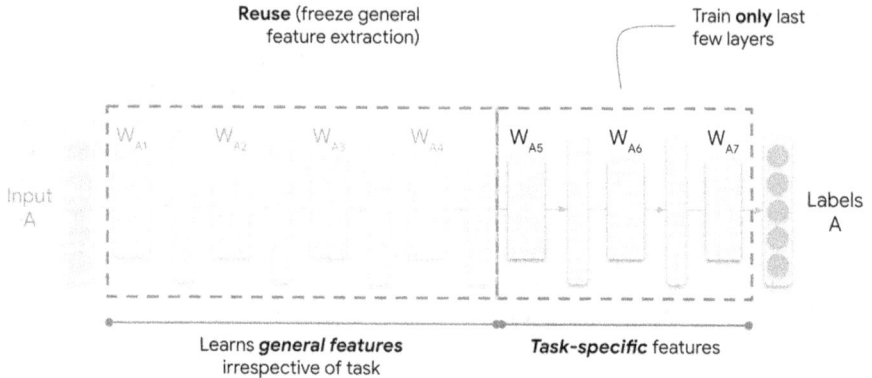

FIGURE 4.7   The pitfalls of transfer learning.

the top, prompting us to omit the topmost classification layers. TensorFlow simplifies this process with a parameter, include_top=False, which allows us to initiate a MobileNetV2 model armed with pretrained ImageNet weights and devoid of the classification layers at the top, making it ideally suited for feature extraction.

After establishing our base model, we employ transfer learning [48,49], a technique that repurposes a model created for one task to adapt it for another, albeit related, task. By leveraging existing models, we not only conserve computational energy but also enhance the model's performance with limited data, which might otherwise lead to overfitting if the model were trained from scratch (see Figure 4.7). Thus, transfer learning offers three key advantages: it reduces energy consumption, accelerates model convergence, and increases the potential for higher accuracy. However, the implementation of transfer learning can be challenging, and its success is not always assured.

4.4.2.4.2   Add a Classification Head

The conventional MobileNetV2 architecture was customized and refined with the integration of three additional layers. These enhancements comprised a 7×7 average pooling layer, a dropout layer with a rate of 0.2, and a dense_1 layer for detection. Excluding these newly added layers, the rest of the architecture was employed for feature extraction. The detailed structure of each layer, including their output dimensions, shape, and the number of parameters, is described in Table 4.1 for more details. In our tailored MobileNetV2 model, there were 1,281 parameters available for training, whereas 2,257,984 parameters were non-trainable, making up a total of 2,259,265 parameters. In contrast, the original MobileNetV2 model was designed with 2,257,984 non-trainable parameters, without any parameters allocated for training, summing up to 2,257,984 parameters in total. The primary objective of our enhanced model was to minimize the quantity of trainable parameters relative to the original MobileNetV2. This is because it was unnecessary to train additional parameters when utilizing the pre-established layers for feature extraction, which is crucial in the identification of faces with or without masks.

TABLE 4.1

Architecture of the "model_2"

Layer (Type)	Output Shape	Param #
input_6 (InputLayer)	(None, 224, 224, 3)	0
sequential_1 (Sequential)	(None, None, None, 3)	0
tf.math.truediv_2 (TFOpLambda)	(None, 224, 224, 3)	0
tf.math.subtract_2 (TFOpLambda)	(None, 224, 224, 3)	0
mobilenetv2_1.00_224 (Functional)	(None, 7, 7, 1280)	2,257,984
global_average_pooling2d_1 (GlobalAveragePooling2D)	(None, 1280)	0
dropout_2 (Dropout)	(None, 1280)	0
dense_1 (Dense)	(None, 1)	1281

Total params: 2,259,265 (8.62 MB)

Trainable params: 1281 (5.00 KB)

Non-trainable params: 2,257,984 (8.61 MB)

4.4.3 MODEL COMPRESSION

This section focuses on the critical process of model conversion, specifically tailoring the enhanced MobileNetV2 for deployment on resource-constrained devices typically encountered in TinyML applications. The transformation of the model into a TensorFlow Lite format is crucial, as it facilitates its execution within the limited computational environments characteristic of edge devices. This conversion is achieved through two principal compression techniques: quantization to eight-bit integers and model pruning [50–55]. Each technique contributes uniquely to the reduction of the model's size and computational demand without significantly compromising performance, a balancing act of utmost importance in TinyML.

4.4.3.1 TensorFlow Lite Model Generation

In the context of deploying machine learning models on resource-constrained devices, TensorFlow Lite [56] serves as a powerful tool that facilitates the conversion and optimization of TensorFlow models into a format that is more efficient for mobile and embedded devices. Below, we'll delve into two key techniques used in this process: Quantization and Pruning.

Dynamic Range Quantization Involves converting floating-point numbers to integers. The formula for dynamic range quantization is given by

$$q = \text{round}\left(\frac{r}{S}\right) + Z, \tag{4.8}$$

where

- r is the real (floating-point) value.
- S is the scale factor.

- Z is the zero-point, representing the real number zero in the quantized space.
- q is the quantized integer value.

The reverse conversion from quantized value back to floating-point is

$$r \approx S \cdot (q - Z), \tag{4.9}$$

The scale factor S is calculated as

$$S = \frac{\text{Range of } r}{\text{Range of } q}, \tag{4.10}$$

Pruning reduces the model size by setting the less significant weights to zero. The pruning condition is defined as

$$W_{ij}' = \begin{cases} 0 & \text{if } W_{-}\{ij\} < \theta \\ W_{ij} & \text{otherwise} \end{cases}, \tag{4.11}$$

where

- W is the original weight matrix.
- W' is the pruned weight matrix.
- θ is the pruning threshold.

The sparsity level of the pruned matrix is

$$\text{Sparsity} = \frac{\text{Number of zero elements in } W'}{\text{Total number of elements in } W}. \tag{4.12}$$

Both quantization and pruning are essential techniques for optimizing machine learning models for deployment on resource-constrained devices. Quantization effectively reduces the computational resources required by converting floating-point values to integers, which accelerates inference and reduces model size. Pruning increases model efficiency by eliminating weights that have minimal impact on the model's output, thereby enhancing computational speed and potentially improving model generalization. Together, these techniques enable the deployment of advanced machine learning models in environments where computational resources, memory, and power are limited.

4.4.3.2 Generate a TensorFlow Lite for Microcontrollers Model

After the initial transformation into TensorFlow Lite format, the model is further modified to work with the TensorFlow Lite Micro (TFLM) runtime [57]. TFLM is tailored for use in microcontrollers and similar devices with minimal memory capacity. The adaptation involves translating the TFLite model into a C source file, which includes an array of bytes that represent the model. This conversion can be achieved using the "*xxd*" command-line utility or through a Python script provided by TensorFlow.

4.5   EXPERIMENTAL ANALYSIS AND EVALUATION

This section delves into the comprehensive evaluation of our TinyML-based mask detection model. We outline the experimental setup, methodologies employed for testing, and a detailed analysis of the model's performance across various metrics. By systematically examining the results, we assess the efficacy of the MobileNetV2 architecture when adapted for real-time mask detection on constrained devices.

4.5.1   EXPERIMENT

The experimental setup, outlined in Table 4.2, was carefully designed to test the performance of the TinyML-based mask detection model within a realistic yet controlled environment. Equipped with a high-specification Dell PC and an array of sophisticated development tools, including Python, TensorFlow, Keras, and OpenCV, this environment facilitated the development and testing of complex image processing algorithms. The OV7670 camera module (see Figure 4.8), used in conjunction with the Arduino Nano 33 BLE microcontroller, formed a robust image acquisition system essential for capturing real-time data.

The Arduino Nano 33 BLE was specifically utilized to deploy our TinyML model (see Figure 4.9). During the experiments, if the model detected that a person was wearing a mask, the LED on the Arduino turned green, indicating compliance.

TABLE 4.2
Execution Environment Specifications

Component	Specification
PC	Brand: Dell
	CPU: Intel(R) Core(TM) i5-8400H CPU @ 2.50GHz
	RAM: 8GB
	Storage: 500GB SSD
	OS: Windows 10
Development Tools	Python
	TensorFlow
	Keras
	Numpy
	Pandas
	Jupyter Notebook
	Visual Studio Code
	OpenCV
Camera	OV7670
Microcontroller	Arduino Nano 33 BLE
Processor	Cortex-M4 running at 64 MHz
Memory	Flash: 1 MB SRAM: 256 KB

FIGURE 4.8   The OV7670 camera overview.

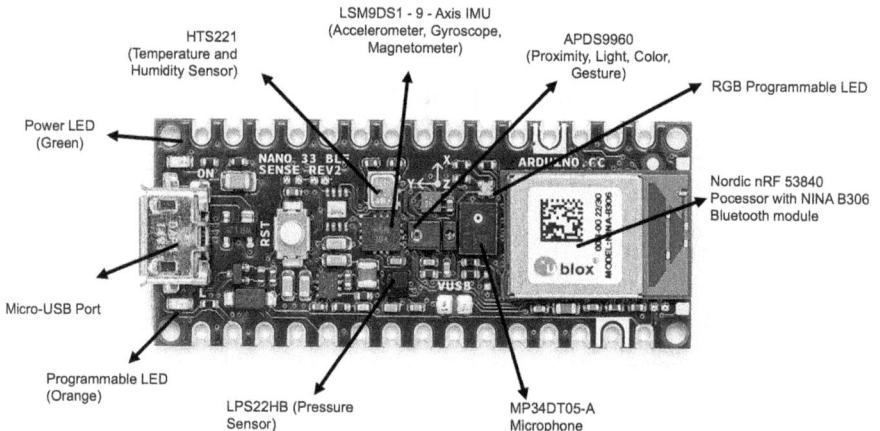

FIGURE 4.9   Arduino Nano 33 BLE sense hardware overview.

Conversely, if no mask was detected, the LED turned red, signaling non-compliance. This setup not only enabled the effective deployment of our advanced TinyML model but also supported a thorough analysis of its performance metrics such as real-time execution capability, accuracy in mask detection, and operational efficiency on the resource-constrained Cortex-M4 processor. The experiment highlights the model's applicability in public health monitoring, particularly in scenarios that demand low power consumption while maintaining high operational performance.

4.5.2   Training Process

In our approach, the foundational layers of the pretrained MobileNetV2 model were kept frozen to leverage the existing learned features, and a new set of trainable layers was introduced. These layers were specifically trained using our dataset to distinguish between individuals wearing masks and those not. This strategy allowed us to utilize the inherent strengths of the MobileNetV2 model, including its pre-tuned bias weights, without the need for additional computational resources and without avoiding the problem of catastrophic forgetting. Subsequently, the model was further refined by introducing additional training parameters: a learning rate of 0.001, a training duration of ten epochs, and a batch size of 32. For optimizing the training process, we employed the Adam optimizer [58], renowned for its efficiency in various deep learning applications, particularly in computer vision. The Adam optimizer dynamically adjusts the learning rate for each weight in the neural network by estimating the gradients' first and second moments. This enabled effective transfer learning, allowing for precise adjustments to the neural network's weights with each batch processed. After completing the training, the neural network was saved, enabling us to make accurate predictions on new images from the dataset. This method ensured that our model was both robust and efficient, ready for deployment in real-world scenarios requiring mask detection.

4.5.3   Evaluation Metrics

In our research, we've used several key evaluation metrics that are essential for assessing the performance of predictive models.

4.5.3.1   Accuracy

We calculate the proportion of real outcomes (including true positives and true negatives) in the total number of instances studied, as shown in equation 4.13:

$$\text{Accuracy} = \frac{\text{Number of Correct Predictions}}{\text{Total Number of Predictions}}. \tag{4.13}$$

4.5.3.2   Precision

We assess the accuracy of positive predictions, focusing on the proportion of positive identifications that were actually correct (see equation 4.14):

$$\text{Precision} = \frac{\text{True Positives}}{\text{True Positives} + \text{False Positives}}. \tag{4.14}$$

4.5.3.3   Recall (Sensitivity)

We measure the ability to find all the relevant cases within a dataset, reflecting the proportion of actual positives correctly identified (see equation 4.15):

$$\text{Recall} = \frac{\text{True Positives}}{\text{True Positives} + \text{False Negatives}}. \tag{4.15}$$

4.5.3.4   F1-Score

We use the harmonic mean of precision and recall to provide a single metric that balances both (see equation 4.16):

$$F1 - Score = 2 \times \frac{Precision \times Recall}{Precision + Recall}.$$

(4.16)

4.5.3.5   Error Rate

We measure the proportion of predictions that were incorrect, highlighting the overall rate of failure in the predictions (see equation 4.17):

$$Error\ Rate = \frac{Number\ of\ Incorrect\ Predictions}{Total\ Number\ of\ Predictions}.$$

(4.17)

4.5.4   RESULTS AND DISCUSSION

In this section, we detail the performance evaluation of our proposed TinyML model designed for real-time mask detection. After training our TinyML model, we undertook a comprehensive evaluation that yielded encouraging outcomes. The model exhibited outstanding performance, achieving an accuracy of 99.6% in the training phase. Remarkably, it reached a perfect accuracy of 100% during the validation phase. These results are graphically represented in Figure 4.10.

Then, we employed confusion matrixes to thoroughly examine our model's ability to accurately distinguish between images of individuals wearing masks and those not wearing them. These matrixes offer a clear visual depiction of our model's classification precision. Figure 4.11 displays both the generalized confusion matrixes and the specific confusion matrix for our proposed model's classifications. We evaluated the proposed Tiny MobileNetV2 model using several metrics, including accuracy, precision, recall (sensitivity), F1-score, and error rate. As illustrated in Figure 4.11, the model demonstrated exceptional performance with a true positive (TP) count of 137, a true negative (TN) count of 138, no false positives (FP), and only one false negative (FN).

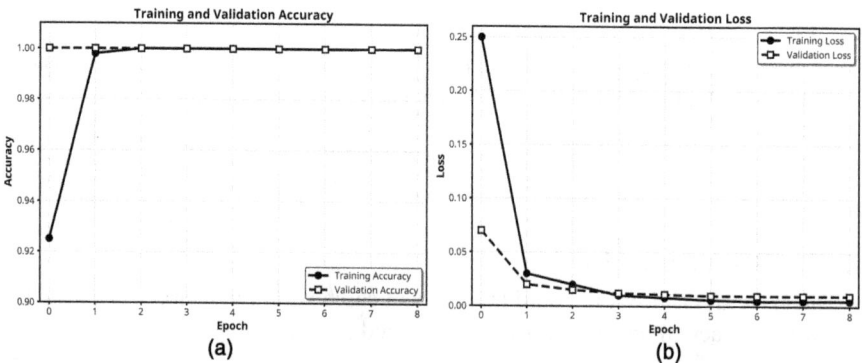

FIGURE 4.10   (a) Accuracy curve for training and validation, (b) Loss curve for training and validation.

Predicted Label   Predicted Label

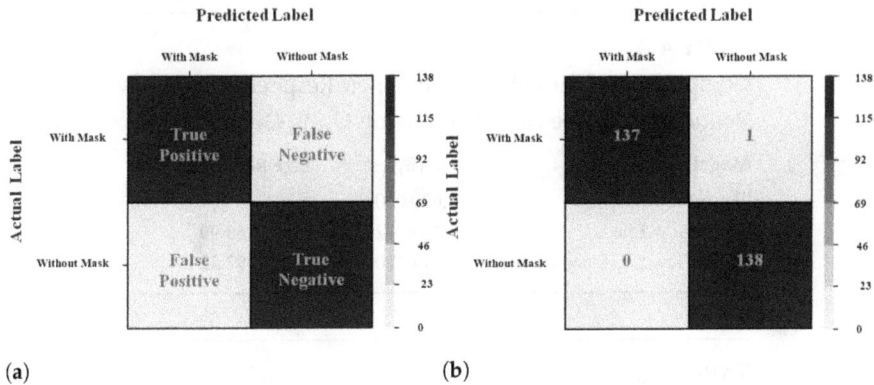

(a)   (b)

FIGURE 4.11 (a) Standard Form Confusion Matrix, (b) Confusion matrix of Tiny MobileNetV2 model.

TABLE 4.3

Comparison of Different Dimension Images of Our Tiny MobileNetV2 Model

Sl. No.	Model Version (Dimension)	Accuracy (%)	Precision (%)	Recall (%)	F1-Score (%)	Error Rate (%)
1	MobileNetV2 (128×128)	97.82	98.53	97.10	97.80	2.18
2	MobileNetV2 (160×160)	98.19	98.54	97.82	98.18	1.81
3	MobileNetV2 (192×192)	98.55	98.55	98.55	98.55	1.45
4	MobileNetV2 (224×224)	99.6	100	99.2	99.64	0.4

In this study, we utilized an image dimension of 224×224 pixels for photo analysis, which significantly enhanced the model's performance metrics. To assess the impact of image dimensions on model effectiveness, we conducted a series of tests. The 224×224 dimension yielded the highest accuracy at 99.6%. Additionally, the specificity metric, which measures the model's ability to correctly identify negatives, also showed remarkable improvement in this dimension. The detailed performance metrics for various image dimensions, including accuracy, precision, recall, F1-score, and error rate, are compared in Table 4.3.

After evaluating all metrics for our Tiny MobileNetV2, we converted the model into a format suitable for deployment on resource-constrained devices. Tables 4.4 and 4.5 detail the size reductions achieved through the application of different optimization techniques. The original TensorFlow model size of 8,620,000 bytes was significantly reduced when converted to TensorFlow Lite, showing a 63.99% decrease in size. Further reductions were achieved with TensorFlow Lite Quantized, which demonstrated a dramatic size reduction of 97.52%. Similarly, the use of pruning techniques also resulted in substantial size reductions. Conversion to TensorFlow Lite with Pruning yielded a model that is 69.33% smaller than the original TensorFlow Lite model, which itself was 40.79% smaller than the original TensorFlow version. We can conclude that the quantization method achieves a higher degree of size reduction

TABLE 4.4

Comparison of Model Sizes and Their Respective Reductions in Size After Conversion Using Quantization

Model	Size (Bytes)	Size Reduction (%)
TensorFlow	8,620,000	—
TensorFlow Lite	3,103,632	63.99
TensorFlow Lite Quantized	214,120	97.52

TABLE 4.5

Comparison of Model Sizes and Their Respective Reductions in Size After Conversion Using Pruning Technique

Model	Size (Bytes)	Size Reduction (%)
TensorFlow	8,620,000	—
TensorFlow Lite	5,103,632	40.79
TensorFlow Lite with Pruning	2,644,120	69.33

TABLE 4.6

Testing the Accuracy of Models Post-Conversion Using Quantization

Model	Testing Accuracy (%)
TensorFlow	99.8
TensorFlow Lite	99.8
TensorFlow Lite with Quantization	99.8

compared to the pruning technique, making it especially effective for deployment in environments with severe size constraints.

The conversion techniques applied to the Tiny MobileNetV2 model, quantization and pruning, have distinct impacts on testing accuracy. Table 4.6 reveals that the model maintains its original testing accuracy of 99.8% across all versions post-quantization, including the TensorFlow original, TensorFlow Lite, and TensorFlow Lite with Quantization. This consistency underscores the effectiveness of quantization in reducing model size without sacrificing accuracy. In contrast, Table 4.7 shows a different outcome when pruning techniques are applied. While the original TensorFlow model retains an accuracy of 99.8%, the TensorFlow Lite version experiences a noticeable drop to 96.4%. Furthermore, the TensorFlow Lite with Pruning version sees a more significant decrease to 92.2%. This suggests that while pruning effectively reduces the model size, it does so at the cost of a marked reduction

TABLE 4.7
Testing the Accuracy of Models Post-Conversion Using Pruning

Model	Testing Accuracy (%)
TensorFlow	99.8
TensorFlow Lite	96.4
TensorFlow Lite with Quantization	92.2

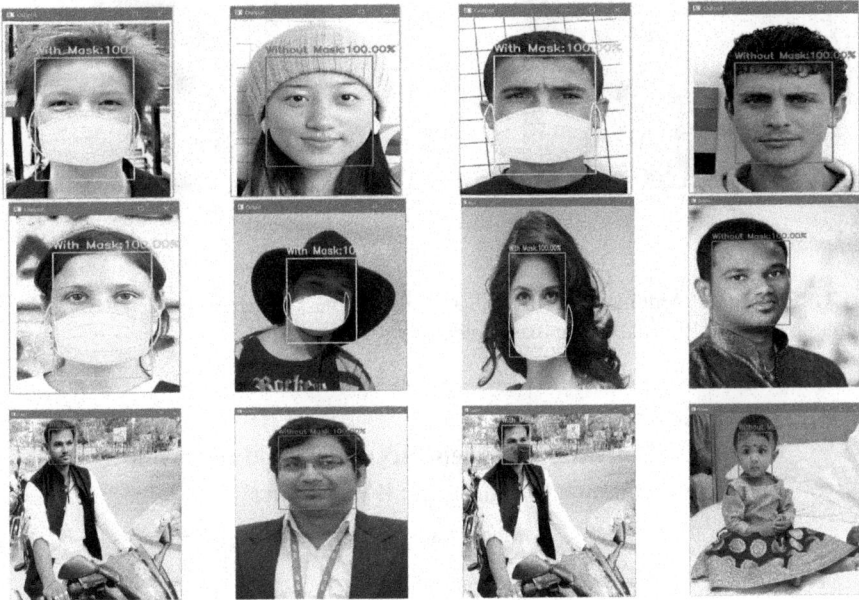

FIGURE 4.12 Evaluating model performance on the test dataset.

in testing accuracy. From these observations, we can conclude that quantization is a superior method for model size reduction when the preservation of accuracy is critical, especially in resource-constrained deployment scenarios. On the other hand, while pruning offers substantial size reduction, it requires careful consideration due to its impact on model accuracy.

Figure 4.12 demonstrates that our model effectively and accurately recognizes images of individuals, whether they are wearing masks or not.

After testing and verifying the satisfactory performance of our model post-conversion, we opted to deploy the quantized TensorFlow Lite model on the Arduino Nano 33 BLE. As previously mentioned, the TensorFlow Lite Quantized model has a compact size of 214 KB, which is compatible with the specifications of the Arduino Nano 33 BLE—featuring a Cortex-M4 processor running at 64 MHz, 1 MB of flash memory, and 256 KB of SRAM. Following deployment on this resource-constrained device, we configured the system such that if the camera detects a person wearing a mask, the LED turns red; conversely, if a person is not wearing a mask, the LED turns green, as demonstrated in Figure 4.13.

The findings from our proposed Tiny MobileNetV2 model, detailed in Table 4.8, underscore its remarkable performance in terms of accuracy while utilizing fewer parameters compared to various other models.

Red LED

(a)

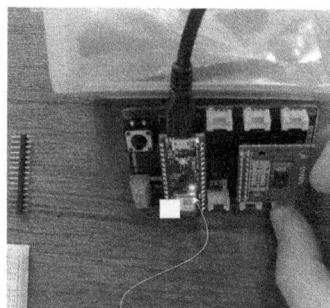

Green LED

(b)

FIGURE 4.13 (a) Arduino Nano 33 BLE showing no mask detected with red LED and (b) Arduino Nano 33 BLE indicating mask detected with green LED.

TABLE 4.8

Comparison of Various Models' Accuracy (bold refers our model performance to distinguish it with others)

Sl. No.	Model Name	Accuracy (%)
1	VGG-Face [59]	68.17
2	OpenFace [59]	63.18
3	DeepFace [59]	63.78
4	MTCNN+FaceNet [60]	64.23
5	FaceNet [59]	67.48
6	IAMGAN [61]	86.50
7	DSA-Face [62]	91.20
8	SSDMNV2 [63]	92.64
9	CBAM [64]	92.61
10	CNNs [65]	91.30
11	LW-CNN [66]	98.47
12	FaceMaskNet21 [67]	88.92
13	GANs [68]	94.10
14	MGL [69]	95.00
15	FaceNet [70]	97.00
16	LPD [71]	97.94
17	CNN [72]	98.00
18	ResNet50 [73]	98.20
19	**Tiny MobileNetV2 (Proposed Model)**	**99.6**

Our results unequivocally demonstrate that the proposed model outperforms existing counterparts.

4.5.4.1 Limitations of Our Proposed Model

While our study on real-time mask detection using TinyML showcases significant advancements, it also encounters certain limitations:

1. **Dataset Limitations**: The model performs exceptionally well with a high-quality, balanced dataset. However, real-world applications often present more diverse and challenging data, which might affect the model's performance. Exploring the model's effectiveness across varied datasets could enhance its robustness and applicability in less controlled environments.
2. **Hardware Constraints**: The study focuses on optimizing model size and computational demands to fit resource-constrained microcontrollers. However, the ongoing advancements in microcontroller capabilities could allow for more complex models without compromising performance, suggesting a need to continually update and optimize model architecture as hardware evolves.
3. **Impact of Quantization and Pruning**: The use of quantization and pruning significantly reduces the model size but may also impact its accuracy and efficiency. A deeper exploration into the trade-offs between model compression and performance is necessary to ensure optimal deployment in real-time applications.

These points highlight the need for ongoing research to address these limitations and improve the technology's efficacy and scope of deployment.

4.6   FEASIBILITY AND COST ANALYSIS OF MODEL IMPLEMENTATION

Implementing the most effective TinyML model for real-time mask detection involves assessing both technical feasibility and cost-effectiveness. This section explores these aspects, aiming to determine a practical approach for widespread deployment.

4.6.1   MODEL EFFICIENCY AND HARDWARE COSTS

The optimized TinyML model, while compact, must be deployed on hardware capable of supporting real-time data processing [74]. The use of microcontrollers like the Arduino Nano 33 BLE, which balances cost and computational capability, presents a cost-effective solution. Advances in hardware may further reduce costs, allowing for the deployment of more sophisticated models without significant financial burden.

4.6.2   SCALABILITY

For large-scale implementation, such as in public transportation systems or retail environments, the scalability of the solution is crucial. The cost per unit decreases with the number of units produced and deployed, making the project more feasible as it scales.

4.6.3 MAINTENANCE AND UPDATES

Continuous operation in diverse environments may require periodic updates to the model and hardware maintenance. Setting up a framework for remote updating of the models and routine maintenance checks can ensure long-term sustainability with minimal additional cost.

4.6.4 INTEGRATION WITH EXISTING SYSTEMS

Integrating this technology with existing surveillance and monitoring systems can reduce implementation costs. By leveraging existing camera systems and infrastructure, the need for extensive new installations is minimized, thereby enhancing the cost-effectiveness of the project.

4.6.5 GOVERNMENT AND INSTITUTIONAL SUPPORT

Securing support from governmental or institutional bodies can significantly reduce financial constraints. Subsidies, grants, or public-private partnerships could provide the necessary funding to cover initial costs, making the deployment feasible in public interest areas.

The combination of technical efficiency, scalable deployment options, and potential institutional support ensures that implementing the optimized TinyML mask detection model can be both feasible and cost-effective, particularly in critical public health contexts.

4.7 CONCLUSION AND FUTURE WORK

This work successfully demonstrates the utility of TinyML in deploying real-time mask detection systems that are efficient and effective, utilizing a highly compressed MobileNetV2 model to achieve an outstanding 99.6% accuracy in mask detection. This approach addresses the crucial need for continuous monitoring of public health safety measures, particularly in managing infectious diseases, and offers a scalable and cost-effective solution to enhance compliance with public health protocols. Our findings not only showcase the capability of TinyML models to operate under constrained resources but also underscore the potential of intelligent systems in public health surveillance. To expand upon this research, future work could focus on several key areas: investigating the model's performance across more diverse environments and datasets to improve its robustness and adaptability to different scenarios, such as varying light conditions and crowds; further testing in real-world settings such as public transport and retail spaces to assess practical deployment challenges and gather user feedback for model refinement; exploring the use of newer and more capable microcontrollers to possibly enhance model performance and efficiency without significant cost increases; studying the integration of the mask detection model with other IoT devices and systems for comprehensive health monitoring solutions; and applying novel machine learning techniques like federated learning for data privacy and model training efficiency. These directions not only promise to enhance the efficacy of public health monitoring systems but also contribute to the broader application of TinyML in healthcare and other critical sectors.

REFERENCES

1. Causadias, J. M., & Neblett, E. W. Jr. (2024). Understanding the impact of the COVID-19 pandemic on the mental health of Latinx children, youth, and families: Clinical challenges and opportunities. *Journal of Clinical Child & Adolescent Psychology*, 53(1), 1–9.
2. Cullen, W., Gulati, G., & Kelly, B. D. (2020). Mental health in the COVID-19 pandemic. *QJM*, 113(5), 311–312.
3. Kauhanen, L., Wan Mohd Yunus, W. M. A., Lempinen, L., Peltonen, K., Gyllenberg, D., Mishina, K., & Sourander, A. (2023). A systematic review of the mental health changes of children and young people before and during the COVID-19 pandemic. *European Child & Adolescent Psychiatry*, 32(6), 995–1013.
4. Kupcova, I., Danisovic, L., Klein, M., & Harsanyi, S. (2023). Effects of the COVID-19 pandemic on mental health, anxiety, and depression. *BMC Psychology*, 11(1), 108.
5. Umbetkulova, S., Kanderzhanova, A., Foster, F., Stolyarova, V., & Cobb-Zygadlo, D. (2024). Mental health changes in healthcare workers during the COVID-19 pandemic: A systematic review of longitudinal studies. *Evaluation & the Health Professions*, 47(1), 11–20.
6. Gupta, S., Sreenivasu, S. V. N., Chouhan, K., Shrivastava, A., Sahu, B., & Potdar, R. M. (2023). Novel face mask detection technique using machine learning to control the COVID'19 pandemic. *Materials Today: Proceedings*, 80, 3714–3718.
7. Himeur, Y., Al-Maadeed, S., Varlamis, I., Al-Maadeed, N., Abualsaud, K., & Mohamed, A. (2023). Face mask detection in smart cities using deep and transfer learning: Lessons learned from the COVID-19 pandemic. *Systems*, 11(2), 107.
8. Liu, Y., Zhang, X., Lin, Y., & Wang, H. (2020). Facial expression recognition via deep action units graph network based on psychological mechanism. *IEEE Transactions on Cognitive and Developmental Systems*, 12(2), 311–322.
9. Xia, Y., Yu, H., Wang, X., Jian, M., & Wang, F.-Y. (2022). Relationaware facial expression recognition. *IEEE Transactions on Cognitive and Developmental Systems*, 14(3), 1143–1154.
10. Abadade, Y., Temouden, A., Bamoumen, H., Benamar, N., Chtouki, Y., & Hafid, A. S. (2023). A comprehensive survey on TinyML. *IEEE Access*, 11, 96892–96922.
11. Giordano, M., Piccinelli, L., & Magno, M. (2022). Survey and comparison of milliwatts microcontrollers for tiny machine learning at the edge. In *2022 IEEE 4th International Conference on Artificial Intelligence Circuits and Systems (AICAS)*, Incheon, Korea, Republic of, 2022.
12. Lamaakal, I., Ouahbi, I., El Makkaoui, K., Maleh, Y., Pławiak, P., & Alblehai, F. (2024). A TinyDL model for gesture-based air handwriting of Arabic numbers and simple Arabic letter recognition. *IEEE Access*, 12, 76589–76605.
13. Lin, J., Zhu, L., Chen, W.-M., Wang, W.-C., & Han, S. (2023). Tiny machine learning: Progress and futures [Feature]. *IEEE Circuits and Systems Magazine*, 23(3), 8–34.
14. Rajapakse, V., Karunanayake, I., & Ahmed, N. (2023). Intelligence at the extreme edge: A survey on reformable TinyML. *ACM Computing Surveys*, 55(13s), 1–30.
15. Lamaakal, I., Essahraui, S., Maleh, Y., El Makkaoui, K., Ouahbi, I., Bouami, M. F., … & Niyato, D. (2025). A comprehensive survey on tiny machine learning for human behavior analysis. *IEEE Internet of Things Journal*, 12, 32419–32443.
16. Wu, C. K., Cheng, C.-T., Uwate, Y., Chen, G., Mumtaz, S., & Tsang, K. F. (2023). State-of-the-art and research opportunities for next-generation consumer electronics. *IEEE Transactions on Consumer Electronics*, 69(4), 937–948.

17. Lamaakal, I., Maleh, Y., El Makkaoui, K., Ouahbi, I., Pławiak, P., Alfarraj, O., ... & Abd El-Latif, A. A. (2025). Tiny language models for automation and control: Overview, potential applications, and future research directions. *Sensors*, 25(5), 1318.
18. Kim, E., Kim, J., Park, J., Ko, H., & Kyung, Y. (2023). TinyML-based classification in an ECG monitoring embedded system. *Computers, Materials and Continua*, 75(1), 1751–1764.
19. Sun, B., Bayes, S., Abotaleb, A. M., & Hassan, M. (2023). The case for tinyML in healthcare: CNNs for real-time on-edge blood pressure estimation. In *Proceedings of the 38th ACM/SIGAPP Symposium on Applied Computing*, Tallinn, Estonia, 629–638.
20. Kasula, B. Y. (2024). Optimizing healthcare delivery: Machine learning applications and innovations for enhanced patient outcomes. *International Journal of Creative Research in Computing*, 6(6), 1–7.
21. Kolasa, K., Admassu, B., Hołownia-Voloskova, M., Kędzior, K. J., Poirrier, J. E., & Perni, S. (2024). Systematic reviews of machine learning in healthcare: A literature review. *Expert Review of Pharmacoeconomics & Outcomes Research*, 24(1), 63–115.
22. Sarker, M. (2024). Revolutionizing healthcare: The role of machine learning in the health sector. *Journal of Artificial Intelligence General Science*, 2(1), 35–48.
23. Bhuiyan, M. N., Rahman, M. M., Billah, M. M., & Saha, D. (2021). Internet of Things (IoT): A review of its enabling technologies in health-care applications, standards protocols, security, and market opportunities. *IEEE Internet of Things Journal*, 8(13), 10474–10498.
24. Haghi, M., Neubert, S., Geissler, A., Fleischer, H., Stoll, N., Stoll, R., & Thurow, K. (2020). A flexible and pervasive IoT-based health-care platform for physiological and environmental parameter monitoring. *IEEE Internet of Things Journal*, 7(6), 5628–5647.
25. Habibzadeh, H., Dinesh, K., Rajabi Shishvan, O., Boggio-Dandry, A., Sharma, G., & Soyata, T. (2020). A survey of healthcare Internet of Things (HIoT): A clinical perspective. *IEEE Internet of Things Journal*, 7(1), 53–71.
26. Islam, M. M., Nooruddin, S., Karray, F., & Muhammad, G. (2023). Internet of Things: Device capabilities, architectures, protocols, and smart applications in the healthcare domain. *IEEE Internet of Things Journal*, 10(4), 3611–3641.
27. Subrahmannian, A., & Behera, S. K. (2022). Chipless RFID sensors for IoT-based healthcare applications: A review of the state of the art. *IEEE Transactions on Instrumentation and Measurement*, 71, 1–20.
28. Howard, A. G., Zhu, M., Chen, B., Kalenichenko, D., Wang, W., Weyand, T., Andreetto, M. and Adam, H. (2017). MobileNets: Efficient convolutional neural networks for mobile vision applications. *arXiv preprint arXiv:1704.04861*.
29. Yadav, S. (2020). Deep learning-based safe social distancing and face mask detection in public areas for COVID-19 safety guidelines adherence. *International Journal of Research in Applied Science and Engineering Technology*, 8(7), 1368–1375. https://doi.org/10.22214/ijraset.2020.30560
30. Hussain, S. A., & Al Balushi, A. S. A. (2020). A real-time face emotion classification and recognition using deep learning model. *Journal of Physics: Conference Series*, 1432(1), 012087.
31. Ejaz, M. S., Islam, M. R., Sifatullah, M., & Sarker, A. (2019). Implementation of principal component analysis on masked and non-masked face recognition. In *Proceedings of the 2019 1st International Conference on Advances in Science, Engineering and Robotics Technology (ICASERT)*, Dhaka, Bangladesh, 1–5.
32. Li, S., Ning, X., Yu, L., Zhang, L., Dong, X., Shi, Y., & He, W. (2020). Multi-angle head pose classification when wearing the mask for face recognition under the COVID-19 coronavirus epidemic. In *2020 International Conference on High Performance Big Data and Intelligent Systems (HPBD&IS)*, Shenzhen, China. 1–5.

33. Qin, B., & Li, D. (2020). Identifying facemask-wearing condition using image super-resolution with classification network to prevent COVID-19. *Sensors*, 20(18), 5236.
34. Din, N. U., Javed, K., Bae, S., & Yi, J. (2020). A novel GAN-based network for unmasking masked faces. *IEEE Access*, 8, 44276–44287.
35. Li, C., Wang, R., Li, J., & Fei, L. (2020). Face detection based on YOLOv3. In *Recent Trends in Intelligent Computing, Communication and Devices: Proceedings of ICCD 2018*, Guangzhou, China. 277–284.
36. Chen, Y., Hu, M., Hua, C., Zhai, G., Zhang, J., Li, Q., & Yang, S. X. (2021). Face mask assistant: Detection of face mask service stage based on mobile phone. *IEEE Sensors Journal*, 21(9), 11084–11093.
37. Venkateswarlu, I. B., Kakarla, J., & Prakash, S. (2020). Face mask detection using MobileNet and global pooling block. In *2020 IEEE 4th Conference on Information & Communication Technology (CICT)*, Chennai, India. 1–5.
38. Loey, M., Manogaran, G., Taha, M. H. N., & Khalifa, N. E. M. (2021). A hybrid deep transfer learning model with machine learning methods for face mask detection in the era of the COVID-19 pandemic. *Measurement*, 167, 108288.
39. Boutros, F., Damer, N., Kirchbuchner, F., & Kuijper, A. (2022). Self-restrained triplet loss for accurate masked face recognition. *Pattern Recognition*, 124, 108473.
40. Deng, H., Feng, Z., Qian, G., Lv, X., Li, H., & Li, G. (2021). MFCosface: A masked-face recognition algorithm based on large margin cosine loss. *Applied Sciences*, 11(16), 7310.
41. Zhong, M., Xiong, W., Li, D., Chen, K., & Zhang, L. (2024). MaskDUF: Data uncertainty learning in masked face recognition with mask uncertainty fluctuation. *Expert Systems with Applications*, 238, 121995.
42. Faruque, M. O., Islam, M. R., & Islam, M. T. (2021). Advanced masked face recognition using robust and lightweight deep learning model. *International Journal of Computer Applications*, 975, 8887.
43. ImageNet. Available online: https://www.image-net.org (Accessed on 11 November 2024).
44. Bhatt, D., Patel, C., Talsania, H., Patel, J., Vaghela, R., Pandya, S., Modi, K., & Ghayvat, H. (2021). CNN variants for computer vision: History, architecture, application, challenges, and future scope. *Electronics*, 10(20), 2470.
45. Kottarathil, P. (2020). *Face Mask Lite Dataset*. Kaggle Data.
46. Bermano, A. H., Gal, R., Alaluf, Y., Mokady, R., Nitzan, Y., Tov, O., et al. (2022). State-of-the-art in the architecture, methods, and applications of StyleGAN. *Computer Graphics Forum*, 41(2), 591–611.
47. Shorten, C., & Khoshgoftaar, T. M. (2019). A survey on image data augmentation for deep learning. *Journal of Big Data*, 6(1), 1–48.
48. Iman, M., Arabnia, H. R., & Rasheed, K. (2023). A review of deep transfer learning and recent advancements. *Technologies*, 11(2), 40.
49. Zhuang, F., Qi, Z., Duan, K., Xi, D., Zhu, Y., Zhu, H., Xiong, H., & He, Q. (2020). A comprehensive survey on transfer learning. *Proceedings of the IEEE*, 109(1), 43–76.
50. Gholami, A., Kim, S., Dong, Z., Yao, Z., Mahoney, M. W., & Keutzer, K. (2022). A survey of quantization methods for efficient neural network inference. In Thiruvathukal, G. K., Lu, Y.-H., Kim, J., Chen, Y., & Chen, B. (eds.). *Low-Power Computer Vision*, pp 291–326. Taylor & Francis Group.
51. Hu, J., Lin, P., Zhang, H., Lan, Z., Chen, W., Xie, K., Chen, S., Wang, H. and Chang, S. (2023). A dynamic pruning method on multiple sparse structures in deep neural networks. *IEEE Access*, 11, 38448–38457.
52. Lamaakal, I., El Makkaoui, K., Ouahbi, I., & Maleh, Y. (2024). A TinyML model for gesture-based air handwriting of Arabic numbers recognition. *Procedia Computer Science*, 236, 589–596.

53. Lamaakal, I., Maleh, Y., Ouahbi, I., El Makkaoui, K., & Abd El-Latif, A. A. (2024). A deep learning-powered TinyML model for gesture-based air handwriting simple Arabic letters recognition. In *Proceedings of the International Conference on Digital Technologies and Applications* (pp. 32–42), Benguerir, Morocco. Springer Nature Switzerland.

54. Lamaakal, I., El Mourabit, N., El Makkaoui, K., Ouahbi, I., & Maleh, Y. (2024). Efficient gesture-based recognition of Tifinagh characters in air handwriting with a TinyDL model. In *Proceedings of the 2024 Sixth International Conference on Intelligent Computing in Data Sciences (ICDS)* (pp. 1–8), Marrakech, Morocco. IEEE.

55. Vadera, S., & Ameen, S. (2022). Methods for pruning deep neural networks. *IEEE Access*, 10, 63280–63300.

56. Contoli, C., & Lattanzi, E. (2023). A study on the application of Tensor-Flow compression techniques to human activity recognition. *IEEE Access*, 11, 48046–48058.

57. Manor, E., & Greenberg, S. (2022). Custom hardware inference accelerator for TensorFlow Lite for microcontrollers. *IEEE Access*, 10, 73484–73493.

58. Kingma, D. P., & Ba, J. (2014). Adam: A method for stochastic optimization. *arXiv preprint arXiv:1412.6980*.

59. Chandra, Y. B., & Reddy, G. K. (2020). A comparative analysis of face recognition models on masked faces. *International Journal of Scientific & Technology Research*, 9(10), 30175–0.

60. Hong, Q., Wang, Z., He, Z., Wang, N., Tian, X., & Lu, T. (2020). Masked face recognition with identification association. In *2020 IEEE 32nd International Conference on Tools with Artificial Intelligence (ICTAI)* (pp. 731–735), Baltimore, MD, USA.

61. Geng, M., Peng, P., Huang, Y., & Tian, Y. (2020). Masked face recognition with generative data augmentation and domain-constrained ranking. In *Proceedings of the 28th ACM International Conference on Multimedia* (pp. 2246–2254), Seattle, WA, USA.

62. Wang, Q., & Guo, G. (2021). DSA-Face: Diverse and sparse attentions for face recognition robust to pose variation and occlusion. *IEEE Transactions on Information Forensics and Security*, 16, 4534–4543.

63. Nagrath, P., Jain, R., Madan, A., Arora, R., Kataria, P., & Hemanth, J. (2021). SSDMNV2: A real-time DNN-based face mask detection system using single-shot multibox detector and MobileNetV2. *Sustainable Cities and Society*, 66, 102692.

64. Li, Y., Guo, K., Lu, Y., & Liu, L. (2021). Cropping and attention-based approach for masked face recognition. *Applied Intelligence*, 51, 3012–3025.

65. Hariri, W. (2022). Efficient masked face recognition method during the COVID-19 pandemic. *Signal, Image and Video Processing*, 16(3), 605–612.

66. Farman, H., Khan, T., Khan, Z., Habib, S., Islam, M., & Ammar, A. (2022). Real-time face mask detection to ensure COVID-19 precautionary measures in the developing countries. *Applied Sciences*, 12(8), 3879.

67. Golwalkar, R., & Mehendale, N. (2022). Masked-face recognition using deep metric learning and FaceMaskNet-21. *Applied Intelligence*, 52(11), 13268–13279.

68. Li, C., Ge, S., Zhang, D., & Li, J. (2020). Look through masks: Towards masked face recognition with de-occlusion distillation. In *Proceedings of the 28th ACM International Conference on Multimedia* (pp. 3016–3024), Seattle, WA, USA.

69. Wang, Z., Huang, B., Wang, G., Yi, P., & Jiang, K. (2023). Masked face recognition dataset and application. *IEEE Transactions on Biometrics, Behavior, and Identity Science*, 5, 298–304.

70. Anwar, A., & Raychowdhury, A. (2020). Masked face recognition for secure authentication. *arXiv preprint arXiv:2008.11104*.

71. Ding, F., Peng, P., Huang, Y., Geng, M., & Tian, Y. (2020). Masked face recognition with latent part detection. In *Proceedings of the 28th ACM International Conference on Multimedia* (pp. 2281–2289), Seattle, WA, USA.

72. Goyal, H., Sidana, K., Singh, C., Jain, A., & Jindal, S. (2022). A real-time face mask detection system using convolutional neural network. *Multimedia Tools and Applications*, 81(11), 14999–15015.

73. Sethi, S., Kathuria, M., & Kaushik, T. (2021). Face mask detection using deep learning: An approach to reduce the risk of Coronavirus spread. *Journal of Biomedical Informatics*, 120, 103848.

74. Sabovic, A., Aernouts, M., Subotic, D., Fontaine, J., De Poorter, E., & Famaey, J. (2023). Towards energy-aware TinyML on battery-less IoT devices. *Internet of Things*, 22, 100736.

5 TinyML for Smarter Healthcare

Compact AI Solutions for Medical Challenges

*Chaymae Yahyati, Khalid El Makkaoui,
Ibrahim Ouahbi, and Yassine Maleh*

5.1 INTRODUCTION

TinyML [1] is a unique method that the scientific community suggests using to build safe, self-sufficient devices that can collect, analyze, and alert without disclosing information to third parties. This study looks at the potential solutions that TinyML, an emerging technology that necessitates the integration of machine learning algorithms, may offer for wearable technology and healthcare applications at the edge. It also talks about how TinyML could enhance neural networks to provide gadgets in fields like autonomy and healthcare intelligence [2,3]. In addition to the numerous potential uses of TinyML in healthcare, these gadgets have sparked serious worries about the confidentiality of the personal information that these services gather and retain.

Microcontrollers with limited computational resources and memory enable TinyML to deploy efficient machine learning models, making it ideal for healthcare applications where accuracy, low latency, and data privacy are critical. TinyML's potential in healthcare spans diverse areas, beginning with wearable devices like fitness trackers and biosensors that monitor health metrics such as heart rate, blood pressure, and oxygen levels while triggering real-time emergency alerts. Beyond wearables, TinyML facilitates remote patient monitoring, empowering healthcare providers to track chronic conditions like diabetes or hypertension in real-time. It also enhances diagnosis and treatment by analyzing medical history, symptoms, and lab results to deliver precise, personalized care plans. Finally, TinyML accelerates drug discovery through its ability to evaluate molecular structures, drug interactions, and cellular pathways, helping researchers identify novel drug targets and develop more effective therapies [4–7].

5.2 WEARABLE MEDICAL DEVICES

Using TinyML technology, scientists propose a new way to create self-sufficient, secure, and environmentally friendly devices that can collect, evaluate, and sound an alarm without sharing data with the outside world. Neural networks improved with TinyML can be used to make medical equipment intelligent and self-sufficient [8].

DOI: 10.1201/9781003544449-5

TABLE 5.1
Wearable Medical Devices and Their Functions

Device Type	Details
Wearable fitness trackers	Fitbit, Garmin, and Apple Watch measure exercise activities such as steps taken, distance traveled, calories burned, and heart rate. They frequently include GPS, sleep tracking, and exercise monitoring.
Smartwatches	Combine smartphone notifications with fitness monitoring functions. Can track heart rate, sleep habits, and physical activity.
Heart rate monitors	Measure and monitor heart rate continually during activity, at rest, or under stress. Provide insights into cardiovascular health.
Blood pressure monitors	Enable convenient and continuous blood pressure monitoring throughout the day. Help track changes and share data with healthcare providers.
Glucose monitors	Allow continuous blood glucose monitoring for diabetes management. Use sensors to measure blood or interstitial fluid glucose levels, transmitting data to smartphones.
ECG monitors	Capture heart's electrical activity (e.g., Apple Watch Series 4+). Can detect abnormal heartbeats and conditions like atrial fibrillation.
Sleep trackers	Measure sleep patterns, duration, quality, and phases. Identify sleep disruption, efficiency, and potential disorders.
Posture correctors	Use sensors to detect body position and provide feedback to maintain proper posture, helping prevent poor postural habits.
UV exposure monitors	Measure ultraviolet radiation exposure and provide real-time alerts to help prevent skin damage.
Smart clothing	Integrate sensors into fabrics to track vitals like heart rate, breathing, temperature, and activity levels. Used for remote patient monitoring and athletic performance.

Numerous AI and communication strategies that are appropriate for the upcoming generation of wearable technologies are reviewed in Ref. [9]. The author of [10] has created a gadget called BandX that uses wearable sensors to identify human activity with the aid of TinyML. The popularity of wearable healthcare gadgets has grown in recent years, as they can measure various medical metrics. The Table 5.1 shows wearable medical devices and their functions.

The COVID-19 epidemic has caused a sharp rise in the number of people using wearable technology between 2019 and 2023. The trend in the use of wearable devices is depicted in Table 5.2.

The market for wearable healthcare technology is expanding, and as it matures, more wearable technology will be available to businesses and individuals. An Insider Intelligence analysis projects that by 2023, there will be 91.3 million users of fitness and health applications, up from 88.5 million in 2022.

While wearable device users prioritize easy access to their health data [5,11], several key factors influence adult adoption of healthcare devices: user-friendly design is essential for seamless interaction, alongside robust privacy protections and data security measures to maintain confidentiality; personalized care options significantly enhance engagement, as do devices with straightforward maintenance requirements; affordability

TABLE 5.2

User Growth (in Millions) by Year

Year	Users (Millions)
2019	68.5
2020	87.7
2021	85.2
2022	88.5
2023	91.3

FIGURE 5.1   The widespread use of TinyML in healthcare.

plays a crucial role in accessibility, while effective coordination with healthcare providers and availability of health coach support further determine successful long-term usage.

Microcontrollers, which are commonly used in wearable electronics, are examples of low-resource devices that TinyML employs machine learning models in Refs. [12–14]. Without requiring cloud-based services, TinyML enables machine learning algorithms to run internally on these devices, enabling real-time data processing and inference. The author of [15] created a wearable gadget to stop the spread of contagious diseases using cutting-edge TinyML technology. Figure 5.1 illustrates the widespread use of TinyML in healthcare, including COVID-19 monitoring through wrist-worn devices that track vital signs, hearing aids enhanced with neural networks for real-time speech augmentation, home healthcare systems that assist patients and caregivers with remote monitoring, and eHealth frameworks designed to streamline digital healthcare services.

5.3 REMOTE PATIENT MONITORING

5.3.1 Remote Patient Monitoring: Overview

An RPM system is a medical device that remotely monitors patients' health using TinyML technology. A subfield of machine learning known as "tinyML" applies machine learning models to devices with limited resources, such as microcontrollers, which are tiny, low-power computing devices. RPM, which is commonly used following these steps:

1. **Signal Processing**: Raw physiological data is cleaned, normalized, and transformed into meaningful features for analysis.
2. **AI Methodology**: Machine learning models are trained on processed signals to identify patterns and extract actionable insights.
3. **Diagnose and Treatment**: The AI system provides clinical decision support by correlating model outputs with medical knowledge to suggest diagnoses and therapeutic options.

These are tiny, wearable devices that collect essential health data using a variety of sensors. Heart rate monitors, blood pressure monitors, temperature sensors, and accelerometers are a few examples. These sensors gather real-time physiological data from the patient. The TinyML Healthcare Monitoring Process is as follows:

1. **Data Preprocessing**: Sensors collect raw health data which is cleaned, transformed, and optimized for analysis by removing noise and extracting key features.
2. **TinyML Model Processing**: Resource-efficient machine learning models analyze the processed data directly on devices to identify health patterns and vital signs.
3. **Real-Time Health Insights**: The models generate instant health assessments and automatically alert caregivers when detecting critical conditions or abnormalities.
4. **Secure Data Handling**: Analyzed data is encrypted and transmitted to cloud platforms for remote access by medical teams and long-term health tracking.
5. **Healthcare Delivery**: The system enables continuous monitoring, reduces hospital visits, and improves care accessibility—especially for chronic patients and rural populations.

5.3.2 Remote Patient Monitoring: Challenges

Implementing remote patient monitoring systems presents several key challenges: establishing robust technical infrastructure with reliable connectivity, secure data transfer, and healthcare system interoperability proves particularly difficult in resource-limited areas; ensuring HIPAA-compliant data security and privacy during collection, transmission and storage requires significant technical and financial resources; user adoption barriers exist among elderly or technologically-averse patients who struggle with device interfaces; healthcare workflows require adaptation to effectively incorporate RPM data into clinical decision-making and care coordination; and inconsistent reimbursement policies and regulatory frameworks across regions create financial and administrative hurdles for sustainable implementation.

To overcome these challenges, a collaborative approach uniting healthcare providers, tech developers, and policymakers is essential. By advancing innovative solutions and addressing current limitations, RPM can transform healthcare delivery—enhancing patient outcomes, expanding access to quality care, and creating more efficient health systems.

5.3.3 Remote Patient Monitoring: Potential Benefits

Remote Patient Monitoring (RPM) powered by TinyML transforms healthcare delivery by enabling continuous real-time tracking of critical health metrics like blood pressure, glucose levels, and heart rhythm, allowing clinicians to detect and address abnormalities instantly. This technology alleviates pressure on healthcare systems through early intervention while making quality care accessible to patients in remote locations by eliminating geographical barriers. By giving patients direct access to their health data, TinyML-enhanced RPM fosters greater engagement in personal health management and adherence to treatment plans. Most importantly, its edge computing capability processes data directly on devices to generate personalized insights, creating tailored treatment approaches that account for each patient's unique physiological patterns and needs.

5.4 ROLE IN DIAGNOSIS AND TREATMENT BY HEALTHCARE PROVIDERS

TinyML is revolutionizing healthcare by embedding machine learning directly into low-power medical devices and wearables. As one of healthcare's most promising technological advancements [16], this approach moves intelligent diagnostics from the cloud to the point of care through efficient edge computing. By processing data locally on microcontrollers, TinyML enables real-time health monitoring while maintaining patient privacy—detecting abnormalities faster than traditional methods and generating personalized treatment insights. Clinicians gain portable AI assistance for immediate decision-making, while healthcare systems benefit from reduced latency and infrastructure costs. The technology's ability to operate on minimal power makes quality care accessible even in resource-constrained environments, truly democratizing advanced medical analytics.

Healthcare providers serve as the critical link between patient data and AI-powered clinical decisions. They receive and preprocess diverse medical inputs—from imaging scans to physiological signals—preparing them for TinyML analysis. These optimized machine learning models run directly on microcontrollers at the point of care, enabling real-time disease detection and treatment suggestions. By interpreting the model's diagnostic predictions alongside their medical expertise, clinicians can develop precise, personalized care plans. TinyML thus augments clinical decision-making while maintaining the provider's central role in patient care.

TinyML enables at-home blood pressure monitoring through innovative edge computing solutions. Ahmed and Hassan [17] developed an end-to-end system that leverages microcontroller-powered peripherals to analyze ECG and PPG signals, predicting key metrics—systolic (SBP), diastolic (DBP), and mean arterial pressure (MAP)—directly on-device. By deploying TinyML at the extreme edge, their solution operates independently of cloud infrastructure, enabling continuous monitoring even without network connectivity. Tested on 12,000 ICU cases spanning 500 hours, the system delivers accuracy comparable to server-based alternatives despite strict memory, power, and computational constraints.

To enhance COVID-19 patient monitoring in emergency settings, Fyntanidou et al. [18] developed a wrist-worn device that tracks critical vital signs including body temperature, respiratory rate, and SpO_2. The wearable incorporates TinyML technology to process respiratory analysis locally via an embedded neural network, eliminating cloud data transmission while ensuring patient privacy and data security.

Recent research [19] demonstrates how TinyML is transforming healthcare through its versatile framework capable of integrating diverse machine learning approaches, dynamically selecting or adapting models, optimizing clinical decision-support systems, and continuously improving through adaptive learning. This innovative technology serves multiple eHealth domains—from mental health monitoring to symptom tracking, body scanning, and hygiene management—delivering intelligent solutions at the point of care.

A study [20] presents an innovative TinyML solution leveraging TensorFlow Lite for personalized home healthcare. This system supports rehabilitation patients, manages chronic/acute conditions, and aids caregiver well-being during high-stress scenarios like COVID-19 outbreaks by enabling intelligent, on-device health monitoring.

Rashid et al. [21] developed Tiny RespNet, an efficient convolutional neural network for respiratory symptom analysis. Implemented on the Xilinx Artix-7 100T FPGA with parallel processing capabilities, the system achieves 245 mW power consumption while offering 4.3× better energy efficiency than conventional solutions. When deployed on NVIDIA Jetson TX2 SoC, it analyzes multiple inputs including audio recordings, speech patterns, and demographic data. Validated across three datasets (ESC-50, FSDKaggle2018, CoughVid), the framework accurately detects cough-related respiratory symptoms and dyspnea, demonstrating robust performance in multimodal healthcare environments.

TinyML enables AI deployment on medical wearables and edge devices but faces key hurdles:

(1) Resource constraints—limited processing power, memory (<1 MB), and energy (~245 mW) demand ultra-efficient algorithms; (2) Power management—varying peripheral designs complicate energy optimization; (3) Cost barriers—scalability issues persist despite low per-unit costs; (4) Data security—strict privacy regulations (*HIPAA/GDPR*) govern sensitive health data; (5) Model interpretability—black-box decisions hinder clinical trust; and (6) Data scarcity—small/imbalanced datasets limit model robustness. Addressing these requires co-development of optimized hardware, explainable AI techniques, and cross-disciplinary collaboration between clinicians, engineers, and regulators.

5.5 DRUG DISCOVERY

TinyML is emerging as a powerful enabler of AI-driven pharmaceutical research, particularly in predicting molecular bioactivity and physicochemical properties—an area where academia and industry are making significant joint advances [22]. By processing vast molecular datasets directly on edge devices, this technology accelerates the identification of promising drug candidates through real-time molecular property prediction. Its ability to perform high-throughput virtual screening and optimize lead compound selection locally could dramatically streamline the drug development pipeline, potentially bringing new therapeutics to market faster while maintaining rigorous computational chemistry standards.

TinyML accelerates pharmaceutical development by enabling powerful predictive analytics on molecular datasets. These edge-deployed models assess drug candidate efficacy and safety profiles, significantly reducing reliance on costly laboratory testing. The implementation follows an iterative optimization cycle that progressively refines prediction accuracy throughout the drug discovery pipeline [23].

1. **Data Collection**: Gather molecular data, compound properties, and biological assay results to train the model.
2. **Feature Extraction**: Process data to extract key features (e.g., molecular descriptors, chemical fingerprints).
3. **Model Training**: Train TinyML models (e.g., neural networks, decision trees) on labeled datasets to predict drug efficacy/toxicity.
4. **Model Optimization**: Optimize models via pruning/quantization for efficient deployment on low-power devices.
5. **Model Deployment**: Deploy on microcontrollers for real-time, on-device predictions.
6. **Prediction and Results**: Evaluate new compounds to prioritize leads or filter out unsafe candidates.

The following are some examples of where TinyML is used in drug discovery.

- **Virtual Screening**: Virtual screening becomes feasible by deploying TinyML algorithms on portable devices. This computational approach enables the rapid screening and prioritization of large compound libraries, swiftly identifying promising molecules.
- **Toxicity Prediction**: Early detection of potential toxicity issues is crucial in drug discovery. TinyML algorithms can be trained on toxicology databases to predict compound toxicity, enabling researchers to filter out hazardous candidates and focus on safer alternatives.
- **Drug Repurposing**: TinyML contributes to exploring new therapeutic applications for existing drugs. Through analysis of extensive datasets on drug properties and molecular interactions, TinyML models identify opportunities for repurposing, accelerating the search for new treatments across various diseases.
- **Personalized Medicine**: TinyML algorithms can be employed in personalized medicine approaches. By analyzing individual patient data, including genetic information and health records, TinyML helps identify tailored drug treatments that align with specific patient needs, resulting in more effective and targeted therapies.

Overall, integrating TinyML into drug discovery holds significant potential to enhance efficiency, reduce costs, and expedite the identification of new drugs and treatments. Table 5.3 presents some advantages and challenges of TinyML in drug discovery.

While challenges remain, continuous innovations in TinyML are progressively overcoming these barriers, positioning it as a transformative technology for pharmaceutical research. Its ability to accelerate drug discovery timelines while maintaining cost efficiency makes TinyML a compelling solution for next-generation therapeutics development.

TABLE 5.3
Advantages and Challenges of TinyML in Drug Discovery

Advantages	Challenges
Real-time analysis of complex molecular data accelerates discovery timelines	Limited computational resources constrain model complexity Balancing model size and accuracy requires careful optimization
Low-power implementation reduces infrastructure costs	Large, high-quality training datasets are difficult to obtain
On-device processing eliminates cloud dependence, improving privacy Mobile deployment enables virtual screening anywhere	Black-box nature complicates regulatory approval
Reduces wet-lab experiments through better candidate prioritization	Generalization across biological contexts remains difficult

---

5.6 ETHICS AND TINYML

TinyML enables machine learning on resource-constrained edge devices like microcontrollers and IoT systems. While this technology unlocks significant potential for real-time intelligent applications, its implementation raises critical ethical challenges that warrant careful examination, including:

- **Privacy Risks**: Sensitive data collected by resource-constrained devices may lack robust encryption, increasing vulnerability to breaches.
- **Algorithmic Bias**: Training data biases can perpetuate discrimination, requiring fairness-aware model design.
- **Informed Consent**: Users must understand data collection purposes, usage, and risks, particularly in healthcare and smart environments.
- **Security Vulnerabilities**: Limited compute resources make devices prone to attacks, risking unauthorized access or misuse.
- **Explainability**: Complex models hinder transparency, challenging accountability and trust in decision-making.
- **Sustainability**: Energy-efficient designs and e-waste management are crucial given battery-powered constraints.
- **Data Ownership**: Clear governance is needed to ensure user control over collected data and derived insights.
- **Equity & Access**: Benefits must be accessible across socioeconomic and geographic divides.
- **Regulatory Gaps**: Developing ethical frameworks and accountability mechanisms remains urgent.
- **Trade-Offs**: Balancing accuracy, efficiency, and privacy demands careful ethical consideration.

To ensure the responsible integration of TinyML technologies, collaborative efforts among developers, policymakers, and stakeholders are essential. Proactive engagement in addressing ethical challenges—through transparent design practices, inclusive governance frameworks, and ongoing risk assessment—will foster trustworthy deployments that align with societal values while advancing edge AI capabilities. TinyML enables transformative healthcare applications through real-time monitoring and predictive analytics, yet its implementation requires careful ethical consideration. Key challenges include:

1. **Data Security**: Sensitive health data demands robust encryption and access controls to prevent breaches.
2. **Informed Consent**: Patients must understand data usage, benefits, and risks before sharing health information.
3. **Bias Mitigation**: Models must be audited for fairness to prevent discriminatory outcomes across patient groups.
4. **Transparency**: Explainable AI techniques are needed to build trust in model predictions among clinicians.

5. **Clinical Validation**: Rigorous testing and trials are essential before deployment in high-risk medical decisions.
6. **Accountability**: Clear guidelines must define responsibility for AI-assisted diagnostic or treatment choices.
7. **Equitable Access**: Technology deployment should prioritize underserved populations to reduce care disparities.
8. **Regulatory Compliance**: Adherence to standards (e.g., HIPAA) ensures patient rights and ethical model use.
9. **Continuous Monitoring**: Ongoing updates are required to maintain accuracy amid evolving medical knowledge.

5.7 CONCLUSION

While still an emerging technology, TinyML is demonstrating significant promise for revolutionizing patient care through diverse applications. From remote patient monitoring (RPM) to early disease detection and precision medicine, this innovative approach brings AI-powered analytics directly to point-of-care devices.

TinyML is transforming healthcare by enabling real-time vital sign monitoring through wearable devices, reducing hospitalizations while improving patient safety. This technology also enhances treatment plans by analyzing patient-specific data to optimize medication dosing. For medical imaging, TinyML algorithms efficiently process X-rays, CT scans, and MRIs, helping clinicians make faster, more accurate diagnoses [24]. In chronic disease management (e.g., diabetes, asthma), these systems improve treatment adherence and health outcomes through continuous monitoring and personalized feedback. Most critically, TinyML's ability to instantly detect emergencies enables immediate medical intervention, significantly improving care quality [25].

TinyML-powered assistive technologies offer real-time audio guidance to enhance mobility for individuals with disabilities [26]. Beyond accessibility applications, these systems optimize medication management by dynamically adjusting dosages and schedules to improve adherence and minimize adverse effects [27]. However, widespread adoption requires addressing critical challenges around data privacy, security, and ethical implementation. Realizing TinyML's full potential in healthcare demands sustained collaboration between technologists, medical professionals, and policymakers to develop robust frameworks for responsible deployment [28].

REFERENCES

1. I. Lamaakal, S. Essahraui, Y. Maleh, K. El Makkaoui, I. Ouahbi, and M. F. Bouami, "A comprehensive survey on tiny machine learning for human behavior analysis," *IEEE Internet of Things Journal*, vol. 12, pp.32419–32443, 2025. doi: 10.1109/JIOT.2025.3565688.
2. I. Lamaakal, Y. Maleh, K. El Makkaoui, I. Ouahbi, P. Pławiak, O. Alfarraj, and A. A. Abd El-Latif, "Tiny language models for automation and control: Overview, potential applications, and future research directions," *Sensors*, vol. 25, no. 5, p. 1318, 2025.

3. I. Lamaakal, I. Ouahbi, K. El Makkaoui, Y. Maleh, P. Pławiak, and F. Alblehai, "A TinyDL model for gesture-based air handwriting Arabic numbers and simple Arabic letters recognition," *IEEE Access*, vol. 12, pp. 76589–76605, 2024.

4. M. S. Diab and E. Rodriguez-Villegas, "Embedded machine learning using microcontrollers in wearable and ambulatory systems for health and care applications: A review," *IEEE Access*, vol. 10, pp. 98450–98474, 2022.

5. M. Zennaro, B. Plancher, and V. Janapa Reddi, "TinyML: Applied AI for development," in *UN 7th Multistakeholder Forum on Science, Technology and Innovation for the Sustainable Development Goals*, New York, USA. May 2022.

6. F. Sabry, T. Eltaras, W. Labda, K. Alzoubi, and Q. Malluhi, "Machine learning for healthcare wearable devices: the big picture," *Journal of Healthcare Engineering*, vol. 2022, no. 1, p. 4653923, 2022.

7. S. Beniczky, P. Karoly, E. Nurse, P. Ryvlin, and M. Cook, "Machine learning and wearable devices of the future," *Epilepsia*, vol. 62, suppl. 1, pp. S116–S124, 2021.

8. V. Tsoukas, E. Boumpa, G. Giannakas, and A. Kakarountas, "A review of machine learning and TinyML in healthcare," in *Proceedings of the 25th Pan-Hellenic Conference on Informatics (PCI)*, Volos, Greece. Nov. 2021, pp. 69–73.

9. R. Sanchez-Iborra, "LPWAN and embedded machine learning as enablers for the next generation of wearable devices," *Sensors*, vol. 21, no. 15, p. 5218, 2021.

10. B. Saha, R. Samanta, S. Ghosh, and R. B. Roy, "Bandx: An intelligent IoT-band for human activity recognition based on TinyML," in *Proceedings of the 24th International Conference on Distributed Computing and Networking (ICDCN)*, Kharagpur, India. Jan. 2023, pp. 284–285.

11. I. Lamaakal, Z. Charroud, Y. Maleh, I. Ouahbi, and K. E. Makkaoui, "Optimizing breast calcification detection in mammography using PySpark: A big data and machine learning approach," in *Proceedings of the International Conference on Data Analytical Management*, Singapore: Springer, 2025, pp. 617–627.

12. I. Lamaakal, N. El Mourabit, K. El Makkaoui, I. Ouahbi and Y. Maleh, "Efficient gesture-based recognition of tifinagh characters in air handwriting with a TinyDL model," in *2024 Sixth International Conference on Intelligent Computing in Data Sciences (ICDS)*, Marrakech, Morocco, 2024, pp. 1–8, doi: 10.1109/ICDS62089.2024.10756483.

13. I. Lamaakal, Y. Maleh, I. Ouahbi, K. El Makkaoui, and A. A. Abd El-Latif, "A deep learning-powered TinyML model for gesture-based air handwriting simple arabic letters recognition," in *International Conference on Digital Technologies and Applications*, Benguerir, Morocco. Cham: Springer Nature Switzerland, May 2024, pp. 32–42.

14. I. Lamaakal, K. El Makkaoui, I. Ouahbi, and Y. Maleh, "A TinyML model for gesture-based air handwriting Arabic numbers recognition," *Procedia Computer Science*, vol. 236, pp. 589–596, 2024.

15. R. M. Umutoni, M. M. Ogore, R. L. Savanna, D. Hanyurwimfura, J. Nsenga, D. Mukanyirigira, *et al.*, "Integration of TinyML-based proximity and couch sensing in wearable devices for monitoring infectious disease's social distance compliance," in *Proceedings of the 2023 12th International Conference on Software and Computer Applications (ICSCA)*, Kuantan, Malaysia. Feb. 2023, pp. 349–355.

16. R. Sanchez-Iborra and A. F. Skarmeta, "TinyML-enabled frugal smart objects: Challenges and opportunities," *IEEE Circuits and Systems Magazine*, vol. 20, no. 3, pp. 4–18, 2020.

17. K. Ahmed and M. Hassan, "TinyCare: A TinyML-based low-cost continuous blood pressure estimation on the extreme edge," in *Proceedings of the 2022 IEEE 10th International Conference on Healthcare Informatics (ICHI)*, Rochester, MN, USA. pp. 264–275, June 2022.

18. B. Fyntanidou, M. Zouka, A. Apostolopoulou, P. D. Bamidis, A. Billis, K. Mitsopoulos *et al.*, "IoT-based smart triage of COVID-19 suspicious cases in the emergency department," in *Proceedings of the 2020 IEEE Globecom Workshops (GC Wkshps)*, Taipei, Taiwan. pp. 1–6, Dec. 2020.
19. P. K. Padhi and F. Charrua-Santos, "6G enabled tactile internet and cognitive internet of healthcare everything: Towards a theoretical framework," *Applied System Innovation*, vol. 4, no. 3, p. 66, 2021.
20. S. Yamanoor and N. S. Yamanoor, "Position paper: Low-cost solutions for home-based healthcare," in *Proceedings of the 2021 International Conference on COMmunication Systems & NETworkS (COMSNETS)*, Bengaluru, India, Jan. 2021, pp. 709–714.
21. H. A. Rashid, H. Ren, A. N. Mazumder, and T. Mohsenin, "Tiny RespNet: A scalable multimodal TinyCNN processor for automatic detection of respiratory symptoms," 2020.
22. N. Stephenson, E. Shane, J. Chase, J. Rowland, D. Ries, N. Justice, *et al.*, "Survey of machine learning techniques in drug discovery," *Current Drug Metabolism*, vol. 20, no. 3, pp. 185–193, 2019.
23. A. Volkamer, S. Riniker, E. Nittinger, J. Lanini, F. Grisoni, E. Evertsson, *et al.*, "Machine learning for small molecule drug discovery in academia and industry," *Artificial Intelligence in the Life Sciences*, vol. 3, p. 100056, 2023.
24. M. Mahmud, M. S. Kaiser, A. Hussain, and S. Vassanelli, "Applications of deep learning and reinforcement learning to biological data," *IEEE Transactions on Neural Networks and Learning Systems*, vol. 29, no. 6, pp. 2063–2079, Jun. 2018.
25. V. Bhatt and S. Chakraborty, "Improving service engagement in healthcare through internet of things based healthcare systems," *Journal of Science and Technology Policy Management*, vol. 14, no. 1, pp. 53–73, 2023.
26. R. Banerjee and S. Mitra, "Artificial Intelligence (AI) for assistive healthcare: Applications, benefits, and challenges," *ACM Transactions on Accessible Computing*, vol. 14, no. 1, pp. 1–29, 2021.
27. Y. Zhang, Z. Luo, and Z. Hou, "A review of artificial intelligence and machine learning applications in smart drug delivery systems," *Pharmaceutical Research*, vol. 38, no. 5, pp. 983–1000, 2021.
28. J. H. Chen, S. M. Asch, and T. Delbanco, "Unlocking the power of digital health data for better care," *The New England Journal of Medicine*, vol. 381, no. 19, pp. 1798–1801, 2019.

6 Adaptive Energy Modeling and Communication Optimization for LoRaWAN-Based IoT Networks

Ismail Lamaakal, Yassine Maleh,
Khalid El Makkaoui, and Ibrahim Ouahbi

6.1 INTRODUCTION

Internet of Things (IoT) [1] devices hold the potential to simplify daily life and enhance people's well-being both at home and at work, as well as in various environments. With the considerable advancement in communication technology, the IoT concept is gradually being embraced across numerous applications in diverse sectors. Various research efforts have been undertaken to better understand and address the communication challenges associated with these technologies. A rising segment within the IoT landscape revolves around low-power wide-area networks (LPWANs) [2]. LPWAN technologies are categorized into cellular and non-cellular networks, aiming to facilitate extensive communication while ensuring scalable coverage for a broad spectrum of IoT applications across different settings and application categories. In IoT applications, several long-range communication paradigms have been employed, including Sigfox, INGENU, Weightless SIG, DASH7, and LoRa [3]. Among these, LoRa has garnered significant attention in the IoT domain. This technology operates within unlicensed industrial–scientific–medical (ISM) bands. Given that these end devices (EDs) are small and often rely solely on battery power, accurately estimating the energy model of the IoT-based LoRa system is crucial. Consequently, determining the optimal communication settings for LoRa in IoT networks has proven challenging.

LoRa networks involve multiple processing stages for transmitted data, encompassing whitening, channel encoding, interleaving, and modulation. Evaluating the energy efficiency of various parameter selections necessitates careful consideration of the trade-offs between transmission energy and computational energy. This chapter

DOI: 10.1201/9781003544449-6

delineates a comprehensive communication system and energy model for LoRaWAN, an LPWAN technology. Moreover, it presents a detailed mathematical model calculating energy usage for transmission, processing, and sensing functions in LoRa's compact sensors. It encompasses transceiver modeling utilizing LoRa's default algorithm function procedure. Several state-of-the-art works have addressed the subject of energy modeling and transmission parameter selection in LoRa networks [4–6]. However, those works have not considered the overall key parameters influencing the energy consumption of LoRa devices, which potentially leads to overestimating the lifespan of terminal devices and consequently the overall network longevity. For instance, the energy model presented in Ref. [7] disregards energy usage variations considering accessible parameters like the LoRa coding rate (CR), instead using fixed values for computational energy, despite its variability in each setup.

Other publications have introduced adaptive techniques for LoRa parameter tuning [8,9], predominantly relying on transmission energy to identify an appropriate set. With this intention, we present crucial elements to estimate the energy consumption of LoRa end nodes. This chapter comprises two sections: one about the communication system function and one about the energy consumption calculation of LoRa end nodes. The first part delves into the elements of the LoRa transceiver, providing a comprehensive overview of the communication system. This includes operations like channel encoding/decoding, modulation/demodulation, whitening/dewhitening, and interleaving/de-interleaving, considering the impact of each transmission choice on all blocks. The second section details an energy model highlighting the primary factors contributing to energy consumption in the LoRa communication system. This model is centered on chirp spread spectrum (CSS) modulation and spreading factors (SF) to hybrid settings with coded transmissions under the additive white Gaussian noise (AWGN) communication channel. To construct a fully operational LoRa communication system, the model integrates the Hamming channel coding method and the CSS modulation approach, offering intricate insights into the functionality of each system block. Additionally, the provided energy model serves to prevent the energy consumption of a given end node regarding its position in the network.

6.2   LoRa/LoRaWAN COMPONENTS AND ELEMENTS

IoT LPWANs are a type of wireless communication technology that is specially developed for IoT applications that require a long-range, low-power connection. These networks provide broad coverage, allowing devices to send and receive data over vast distances while spending little energy. Additionally, these networks can be identified by their ability to function in low-bandwidth settings, making them ideal for applications involving sporadic data transfer and battery-powered devices. These networks offer a low-cost alternative for connecting many IoT devices across huge geographic areas [2].

There are various LPWA technologies available, each with its own set of benefits and drawbacks. LoRaWAN, narrowband IoT (NB-IoT), and long-term evolution for machines (LTE-M) are some common LPWA technologies [10]. In this section, we highlight some important LoRa/LoRaWAN elements, including how it functions and its architecture elements.

6.3 LoRaWAN

LoRaWAN is an open standard created by Semtech Corporation that acts as a bridge between the proprietary base physical layer and the top levels of communication, as shown in Figure 6.1. LoRaWAN is one of the media access control (MAC) protocols that enable wireless interconnection and time scheduling between end nodes and the network's base station [12]. To connect wirelessly with their gateways (GWs), the nodes use the ALOHA protocol and time division multiple access (TDMA) scheduling access mechanisms. As shown in Figure 6.2, the nodes are densely dispersed, providing a star-of-star network structure. The received data are sent to backend servers via the GWs. Furthermore, depending on the application's needs, these nodes might be classified as class A, B, or C [13]. Class A is the most frequent because it saves energy by opening two short windows to listen for and prepare for the reception of downlink feedback from the GW following an uplink broadcast. Class B, on the other hand, extends the listening windows to notify the GW of the node's waiting time since it stays open even after transmission occurs. Class C is deemed

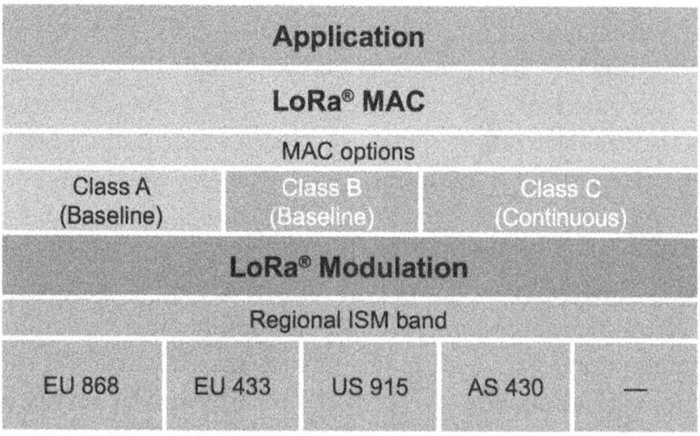

FIGURE 6.1 The LoRaWAN (MAC) protocol stack is implemented on top of LoRa [11].

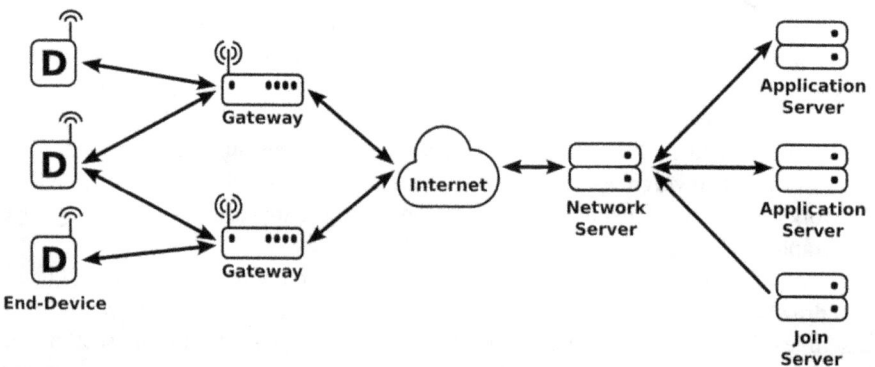

FIGURE 6.2 LoRa network structure.

energy-inefficient since nodes must remain open the majority of the time to provide for continuous and maximum reception windows.

6.3.1   LoRa Physical Layer

LoRa is the physical layer foundation of Semtech's recently introduced long-range LPWA technology. Fundamentally, it employs a patented CSS-derived approach in which many chirps are distributed throughout the occupied frequency spectrum [14]. The basic chirp changes its frequency values instantly by covering the whole related frequency range. The fundamental chirp is incorporated for LoRa by taking several key characteristics into account, including the bandwidth (BW), SFs, CRs, transmission power (TP), and channel signal-to-noise ratio (SNR), which are essential parameters in LoRa communication. As they define different features, the spreading factors (SFs) {7, 8, 9, 10, 11, 12} and various bandwidths (BWs) of 125, 250, and 500 kHz are standard.

LoRa employs the chirp spread spectrum (CSS) modulation technique. Figure 6.3 illustrates that LoRa symbols are modulated across an up-chirp with a bandwidth of 125 kHz. This figure represents an example of using a specific SF and was generated in MATLAB® using the communication chain described in the sections below.

The modulator generates various chirps (up-chirps, down-chirps) and symbol values. In terms of bandwidth, LoRa may send a sample every

$$T_c = \frac{1}{BW},\qquad (6.1)$$

which may also represent the chirp duration. Each sample contains a subset of the information encoded according to the SF bit count. Before modulation, these data are encoded into a non-binary symbol with values in the set {0, 1, 2, ..., $2^{SF} - 1$}.

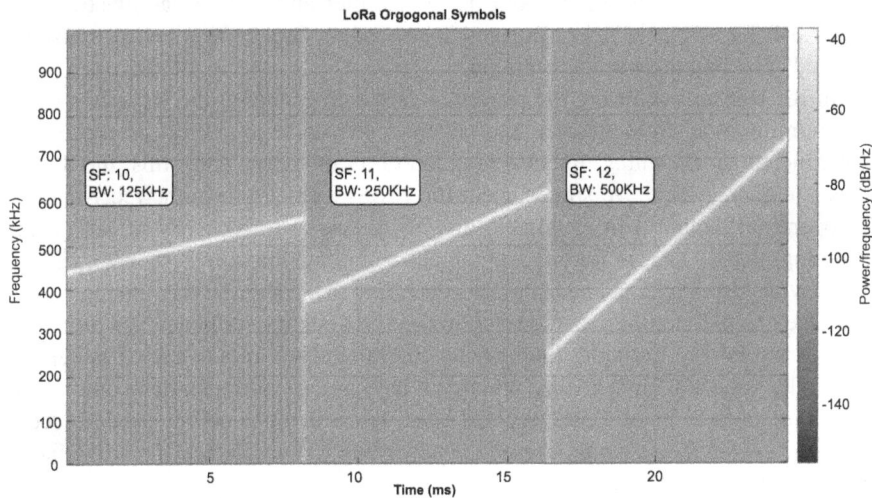

FIGURE 6.3   The description of the SF-12 chirp in the 125 kHz frequency band versus time (ms).

A symbol is transmitted every

$$T\_s = \frac{2^{SF}}{BW},$$ (6.2)

and the higher the SF, the longer it takes to send the symbol.

LoRa incorporates proportional forward error correction (FEC) codes. These codes can group 4-bit data blocks and encode them by adding adjustable parity bits, varying from one to four bits. This flexibility enables different coding rates (CR), where $CR \in \left\{ \frac{4}{5}, \frac{4}{6}, \frac{4}{7}, \frac{4}{8} \right\}$, which are crucial for balancing long-range transmission, low energy usage, and bit rate trade-offs.

Furthermore, the usable bit rate R_b is proportional to the utilized bandwidth, spreading factor, and coding rate and is defined as follows:

$$R_b = \frac{SF \times BW \times CR}{2^{SF}},$$ (6.3)

where $SF \in \{7,8,9,10,11,12\}$, $BW \in \left\{ 125 \text{ kHz}, 250 \text{ kHz}, 500 \text{ kHz} \right\}$
$CR \in \left\{ \frac{4}{5}, \frac{4}{6}, \frac{4}{7}, \frac{4}{8} \right\}$

6.3.2 LoRa Transceiver Architecture and Signal Processing

This section provides a detailed overview of the functional blocks within a LoRa transceiver. It outlines the mathematical operations associated with each processing stage during both transmission and reception modes [15]. LoRa technology relies heavily on physical layer configurations and dedicated processing blocks to enable long-range data transmission. Each successful transmission requires a sequential set of operations performed by both the transmitter and receiver, as illustrated in Figure 6.4. These blocks are responsible for encoding data packets before transmission and decoding them upon reception.

On the transmission side, the process starts with encoding the input bits using a Hamming code. Subsequently, the encoded bits undergo several preprocessing steps, including whitening, interleaving, and Gray indexing. After this, the data is modulated using the chirp spread spectrum (CSS) technique. On the receiver side, the operation begins with estimating and correcting any frequency offsets in the incoming signal. This is followed by signal demodulation, Gray de-indexing, de-interleaving, dewhitening, and Hamming decoding to retrieve the original input data.

These channel coding techniques enhance the system's resilience against transmission errors. For example, interleaving rearranges bits in a specified pattern to reduce the impact of burst errors, although it does not provide error detection or correction on its own. In the LoRa physical layer, interleaving is mainly used to reconstruct the encoded codewords diagonally according to the selected spreading factor (SF) used during CSS modulation.

LoRa enhances transmission robustness by integrating controllable redundancy bits with the transmitted data, especially to combat channel disturbances. This is

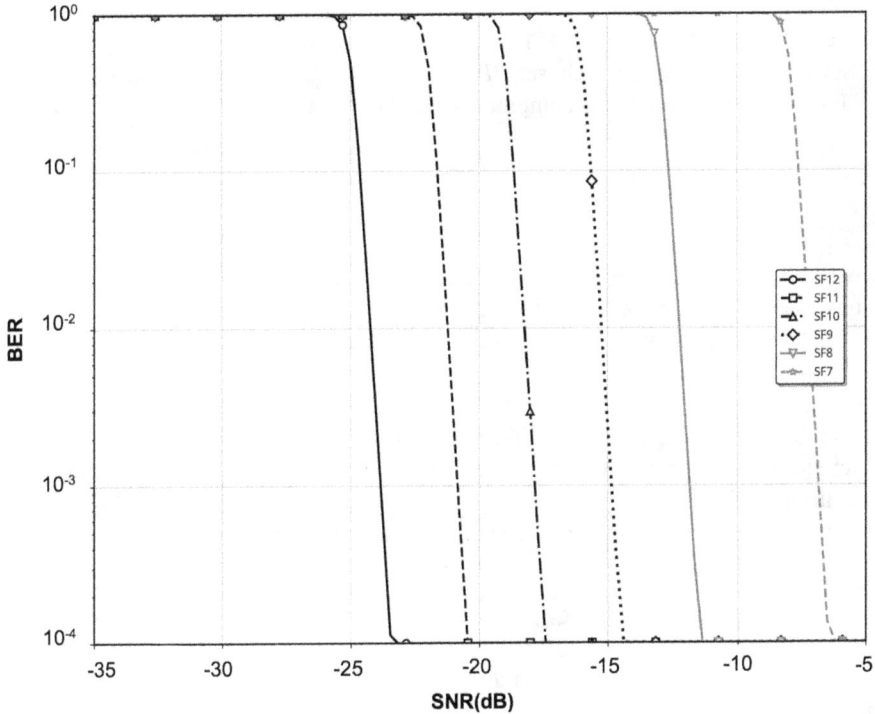

FIGURE 6.4 CSS modulation BER versus SNR performance in an AWGN channel model.

implemented by using a Hamming code $H(k, n)$ with variable codeword lengths governed by the coding rate (CR), defined as

$$CR = \frac{4}{4+\alpha}, \quad \alpha \in \{1,2,3,4\},$$

This yields $CR \in \left\{\frac{4}{5},\frac{4}{6},\frac{4}{7},\frac{4}{8}\right\}$ Thus, each group of four bits is cyclically encoded to form codewords of length $n = 4 + \alpha_{cr}$, where α_{cr} defines the number of parity bits.

The check matrix H of a Hamming (n, k) code is a generator matrix of an orthogonal code C, denoted by C^T. If H is the check matrix, then it is an $(n-k) \times k$ matrix, and its rows are orthogonal to all valid codewords in C. The resulting block of encoded codewords of size $(4+\alpha_{cr}) \times q$ is reordered in preparation for symbol generation using LoRa SFs. To enhance resilience, interleaving is applied to these codewords. The length of the whitened vector must match that of the encoded vector to maintain symbol alignment. Whitening introduces randomness to aid clock recovery at the receiver.

Each codeword generated at the output of the FEC encoder is distributed in time through the interleaving process. This redistributes potential error locations, thereby increasing the probability of correcting them during reception. Although the interleaver scrambles bits in a specific sequence, it does not perform correction or detection itself. In the LoRa PHY layer, it is used to rearrange diagonally encoded codewords according to the chosen *SF*.

Each generated symbol is computed using the formula

$$S_m = \sum_{p=0}^{SF-1} V_{p,m} \cdot 2^p,$$

where S_m is a non-binary symbol with values in the set $\{0, 1, 2, ..., 2^{SF}-1\}$. Consequently, the total symbol stream can be represented as a vector of M elements.

The symbol generation begins with applying Gray indexing to interleaved binary codewords of length *SF*, mapping them to non-binary symbols. For example, if $SF = 7$, then each symbol can take a value from 0 to 127. These symbols are then passed to the CSS modulator. The modulation technique ensures each packet assumes a distinct shape in the time-frequency space, increasing its robustness against channel impairments.

The baseband chirp waveform used in CSS modulation is defined as

$$C_0(t) = A(t) \cdot e^{j\pi \frac{BW}{T_s} t^2 + \phi_0}, \tag{6.4}$$

where $A(t)$ is the envelope function and ϕ_0 is the initial phase. For a symbol S_m, the modulated signal is

$$C_m(t) = A(t) \cdot e^{j2\pi \left(\frac{BW}{T_s} t^2 + \frac{S_m}{2^{SF}} t + \phi \right)}, \tag{6.5}$$

By applying the Shannon sampling theorem and assuming the sampling frequency equals BW, we obtain the discrete form:

$$C_m[k] = e^{j2\pi \left(BW \cdot (kT_c)^2 + S_m \cdot kT_c \right)}, \tag{6.6}$$

The entire transmitted signal comprising all symbols is given by:

$$T_{S_m}(kT_c) = \sum_{m=0}^{M-1} C_S\left(k - m \cdot 2^{SF}\right), \tag{6.7}$$

The received signal over an AWGN block-fading channel is

$$RX[n] = h \cdot T_{S_m}[n] + Z[n], \tag{6.8}$$

where h is the channel gain, and $Z[n]$ is additive white Gaussian noise (AWGN) with zero mean and a certain variance.

The demodulation process begins with de-chirping, where the received signal is multiplied by the complex conjugate of the base chirp:

$$RD[k] = RX[k] \cdot e^{-j\pi BW(kT_c)^2}, \quad \forall k \in \mathbb{Z}. \tag{6.9}$$

The most likely received symbol $\hat{S}\,m$ is then determined by

$$\hat{S}_m = \underset{S_m}{\arg\max} \ \mathrm{Re}\left\{ DFT\left[RD\left(k + m \cdot 2^{SF}\right) \right] \right\}, \tag{6.10}$$

The resulting non-binary symbols are converted back to binary using inverse Gray indexing. Dewhitening is performed by XORing the binary stream with the same pseudo-random sequence used during whitening:

$$\mathrm{DeWhitened}[i] = \mathrm{Recovered}[i] \oplus \mathrm{WhiteningSeq}[i].$$

Next, the data is stored in a rectangular array similar to the transmission side, but now filled row-wise. This array is passed to the decoder, which uses the inverse interleaving to restore the original bit positions. Finally, Hamming decoding is applied to each codeword to correct any bit errors and recover the transmitted data accurately.

6.3.3 LoRa Energy Modeling

This section focuses on the energy modeling of LoRa devices. The provided parts estimate the overall LoRa network energy drain; nevertheless, a full examination must include all power consumption factors in each node.

6.3.3.1 Communication Energy

LoRa nodes operate with limited energy budgets and are typically active for a pre-defined duration T_{active}, after which they enter a deactivated or sleep mode lasting T_{off}. In standby mode, the node maintains radio listening capability with minimal energy consumption, unlike the active mode, where it performs multiple energy-intensive operations. During a complete communication cycle, a node initiates transmission by sending acknowledgment messages embedded in the preamble, followed by payload transmission after establishing communication with the gateway (GW). Consequently, the total transmission duration, denoted T_{packet}, can be expressed as

$$T_{packet} = T_{preamble} + T_{payload}, \tag{6.11}$$

where $T_{preamble}$ and $T_{payload}$ represent the durations of the preamble and the physical payload, respectively. The preamble duration is generally determined by the type of end device, while the payload duration depends on the number of transmitted symbols. These durations are given by

$$T_{preamble} = (4.25 + N_{pr}) \cdot T_s, \tag{6.12}$$

$$T_{payload} = N_{phy} \cdot T_s, \tag{6.13}$$

where N_{pr} is the number of preamble symbols, T_s is the duration of one symbol, and N_{phy} is the number of symbols forming the payload.

The communication module supports both uplink transmission and downlink reception. The RF circuit manages data transmission and reception by the end device (ED). Typically, the transmitting node initiates communication by sending two uplink alerts to the GW at predefined intervals. It then activates two successive listening windows in anticipation of a response. If no signal is received during the first listening window, the node opens a second window. If both windows fail to capture a response, the node waits until the next duty cycle to reattempt transmission. Once both listening windows are exhausted, the node terminates the uplink attempt within the allocated preamble period.

The total energy consumed by the RF circuitry during communication, denoted $E_{RF}(t)$, includes energy for transmission and energy consumed during both listening windows and is expressed as

$$E_{RF}(t) = E_{tx}(t) + E_{rx,w1}(t) + E_{rx,w2}(t), \qquad (6.14)$$

where $E_{tx}(t)$ is the transmission energy, and $E_{rx,w1}(t)$ and $E_{rx,w2}(t)$ represent the energy consumed during reception windows 1 and 2, respectively. The energy during the first listening window is given by

$$E_{rx,w1}(t) = P_{rx,w1} \cdot N_{sym} \cdot T_s, \qquad (6.15)$$

where $P_{rx,w1}$ is the power required during the first window, N_{sym} is the number of symbols exchanged (typically 8 for SF11 and SF12, and also 8 for lower SFs), and T_s is the symbol duration.

The energy consumed during the second window is given by

$$E_{rx,w2}(t) = P_{rx,w2} \cdot \frac{32 + 2^{SF}}{BW}, \qquad (6.16)$$

where $P_{rx,w2}$ is the power required for the second window, SF is the spreading factor, and BW is the transmission bandwidth.

The transmission energy is computed as

$$E_{tx}(t) = P_{tx} \cdot T_{packet}, \qquad (6.17)$$

where P_{tx} is the transmit power and T_{packet} is the complete packet duration as defined in equation (6.11).

The required transmission power P_{tx} can be derived from the Friis transmission model and is expressed as

$$P_{tx} = \rho \cdot \left(\frac{NF}{c}\right) \cdot \left(\frac{4\pi f_c}{c}\right)^2 \cdot \left(\frac{d}{d_0}\right)^\alpha, \qquad (6.18)$$

where ρ is the spectral efficiency, NF is the noise figure, f_c is the carrier frequency, c is the speed of light, d is the distance between the transmitter and the receiver, d_0 is the reference distance (typically 1 m), and α is the path-loss exponent.

6.3.3.2 Computation Energy

In addition to the energy expended by the radio communication components, the node's microcontroller unit (MCU) contributes to the total energy consumption through internal data processing tasks. These tasks include channel coding, whitening, interleaving, Gray indexing, and most notably, modulation using chirp spread spectrum (CSS). The energy required by the processing unit arises primarily from two sources: dynamic energy consumed during logic switching and static energy losses due to leakage currents within the silicon circuitry.

The total time spent by the MCU's physical layer block to complete all the processing stages associated with packet preparation and modulation is estimated as

$$T_{MCU} = T_{sc} + T_{cc} + T_{whi} + T_{int} + T_{gr} + T_{css}, \tag{6.19}$$

where T_{sc}, T_{cc}, T_{whi}, T_{int}, T_{gr}, and T_{css} denote the processing times for source coding, channel coding, whitening, interleaving, Gray indexing, and CSS modulation, respectively.

The total energy consumption of the MCU during the active mode, denoted $E_{MCU}(t)$, is composed of energy due to leakage current and the energy required for logic switching. This is formally given by

$$E_{MCU}(t) = E_{leakage}(t) + E_{MCU,switch}(t), \tag{6.20}$$

where $E_{leakage}(t)$ accounts for the static energy loss due to leakage currents, and $E_{MCU,switch}(t)$ represents the dynamic energy consumed during the logic switching events of the processing sequences.

6.3.3.3 Sensing and Circuitry Energy

The energy consumed by a LoRa node is not limited to communication and computation; the sensing subsystem also contributes significantly to the overall energy budget. The sensor unit is responsible for acquiring physical data from the environment, and it typically includes analog sensors, signal conditioning circuitry, and analog-to-digital (ADC) or digital-to-analog converters (DAC) for proper data interfacing with the digital processing unit.

The total energy consumed during sensing and associated circuit operations is denoted by $E_{sc}(t)$, which is composed of the energy consumed by the circuitry ($E_c(t)$) and the energy consumed by the sensor module itself ($E_s(t)$), expressed as

$$E_{sc}(t) = E_c(t) + E_s(t), \tag{6.21}$$

where $E_c(t)$ accounts for the energy dissipated in the node's supporting electronic circuits, such as voltage regulators, interface circuits, and clock generators, while $E_s(t)$ refers to the energy required for sensor operation, including data acquisition and signal conversion. These components collectively determine the sensing energy footprint of the LoRa node during each active cycle.

6.3.3.4 Network Total Consumed Energy

The total energy consumed in a LoRa-based network over time can be assessed by aggregating the energy expenditures of all participating end devices (EDs). Assume a network made up of N_{node} LoRa end devices (EDs), with each device transmitting data over N_{cycle} consecutive cycles. The overall energy consumed by the network, denoted $E_{total}(t)$, is defined as the sum of the energy dissipated by each node D_{id} over all cycles:

$$E_{total}(t) = \sum_{id=1}^{N_{node}} \sum_{n=1}^{N_{cycle}} E_n(D_{id}), \tag{6.22}$$

where $E_n(D_{id})$ represents the energy expended by the id-th node during its n-th duty cycle. This formulation provides a comprehensive estimate of network-wide energy consumption across a defined operational period.

Furthermore, to predict the residual energy of a given end device at a specific time, one can consider the device's initial energy capacity E_{init} and subtract the cumulative energy consumed over past transmission cycles. The residual energy at time t is estimated by

$$E_{residual}(t) = E_{init} - \sum_{n=1}^{N_{cycle}} E_n(D_{id}), \tag{6.23}$$

where the summation accounts for the total energy expended by node D_{id} until time t. This equation enables the tracking of node energy depletion, facilitating energy-aware scheduling and network lifetime optimization strategies.

6.4 METHODOLOGY

Consider a wide-area network composed of hundreds of densely deployed wireless end devices (EDs). These devices are capable of independently sensing and transmitting observed data to a remote monitoring base station without requiring coordination with one another. Each node utilizes chirp spread spectrum (CSS) modulation individually. To coordinate access to the shared wireless medium and avoid collisions, a time division multiple access (TDMA) slot assignment scheme is employed, which organizes transmissions from EDs to the base station.

Given that these nodes operate on limited battery power, it is crucial to minimize energy consumption while maintaining communication reliability. Each node is strategically deployed in the monitoring region and must autonomously select optimal physical layer parameters that balance communication quality with energy efficiency. LoRa devices must adapt transmission parameters to channel conditions and the distance to the gateway. The selectable parameters include spreading factors $SF \in \{7, 8, 9, 10, 11, 12\}$ and coding rates $CR \in \left\{\frac{4}{5}, \frac{4}{6}, \frac{4}{7}, \frac{4}{8}\right\}$.

To assess the performance of these parameters, it is necessary to determine the distance ranges at which each configuration provides the best energy efficiency and reliability. We propose an adaptive algorithm designed to regulate optimal transmission parameters that ensure both network dependability and energy conservation for each LoRa end device. In this evaluation, the configuration using $SF = 7$ is considered as the baseline for the highest energy consumption level. This reference enables relative energy comparisons.

The evaluation employs the following metric to determine the trade-off between energy saved and energy expended by computation. Specifically, if the energy saved from switching to a more efficient configuration exceeds the computational cost incurred by reconfiguration, then the change is considered beneficial. The energy efficiency ratio E(ED) for a given end device

$$\varepsilon(ED) = \frac{E_{tx}(ED)}{E_{pros}} \times 100\%, \tag{6.24}$$

where $E_{tx}(ED)$ represents the total transmission energy under a given configuration, and E_{pros} is the processing energy associated with selecting and switching to that configuration.

The critical distance d_c is the transition point at which two different spreading factors provide identical transmission energy. That is, for SF_j and SF_i, $\Delta E_{tx} = 0$ implies equal transmission efficiency, and the decision point lies at this boundary. For SF_j, the critical distance corresponds to the maximum range achievable, while for SF_i, it represents the minimum range required. This critical distance $d_c(SF_j)$ is computed using

$$d_c(SF_j) = \left[\frac{E_{tx}(SF_j) \cdot 10^{\frac{-G_{CSS}(SF_j) - NF}{10}}}{T_{packet}(SF_j) \cdot \rho \cdot T \cdot K \cdot BW} \cdot \left(\frac{1}{4\pi f_c} \right)^2 \right]^{\frac{1}{\alpha}}, \tag{6.25}$$

where

- $E_{tx}(SF_j)$ is the transmission energy for SF_j,
- $G_{CSS}(SF_j)$ is the CSS modulation gain,
- NF is the receiver noise figure in dB,
- ρ is the spectral efficiency,
- T is the ambient temperature,
- K is Boltzmann's constant,
- BW is the communication bandwidth,
- f_c is the carrier frequency, and
- α is the path-loss exponent.

This formulation supports the implementation of energy-aware heuristics that can dynamically reconfigure LoRa EDs for optimal operation based on their deployment distance and channel condition. Algorithm 6.1 explains the internal mechanism of adaptive parameter selection under energy constraints.

6.4.1 CASE STUDY AND EXPERIMENTAL SETUP

In this section, we assess the energy consumption, energy efficiency, and transmission reliability of LoRa end devices (EDs) utilizing chirp spread spectrum (CSS) modulation in the context

Algorithm 6.1: Adaptive LoRa Transmission Parameter Selection

1: **Input:** Node position, available SFs $\in$ {7, 8, 9, 10, 11, 12}, CRs $\in \left\{\dfrac{4}{5}, \dfrac{4}{6}, \dfrac{4}{7}, \dfrac{4}{8}\right\}$, and distance

to gateway d

2: **Output:** Optimal transmission configuration (SF, CR)

3: **for** each combination of SF and CR, **do**

4: Calculate transmission energy E_{tx}

5: Calculate packet duration T_{packet}

6: Estimate total energy $E_{total} = E_{tx} + E_{processing}$

7: Check reliability condition for distance d

8: **if** communication fails, **then**

9: **continue** to the next setting

10: **end if**

11: **end for**

12: Select configuration with:

13:    (a) Minimum E_{total} **and**

14:    (b) Meets reliability criteria

15: **if** Multiple configurations satisfy (a) and (b) **then**

16:    Choose one with a minimum T_{packet}

17: **end if**

18: **return** Optimal (SF, CR)

of low-power wide-area networks (LPWANs). Using MATLAB, we conduct comprehensive simulations of an ED placed in a star-of-stars topology. The evaluation explores various propagation scenarios driven by physical layer parameters. Table 6.1 outlines the average values of the channel model parameters and electrical characteristics used in our simulations, selected from real-world LoRa transceiver specifications.

To evaluate the energy consumption accurately, we simulate the entire communication chain, accounting for both coding and modulation operations. Bit error rate (BER) performance is determined by executing multiple transmissions across diverse

TABLE 6.1

Simulation Parameters and Their Values

Parameter	Value
Bandwidth (BW)	125 kHz
Coding rate (CR)	4/5, 4/6, 4/7, 4/8
Spreading factor (SF)	7, 8, 9, 10, 11, 12
Path-loss exponent (α)	3 (urban), 4 (suburban)
Voltage (transceiver)	3.3 V
Carrier frequency (f_c)	858 MHz
Sensor unit voltage	2 V
Processing unit voltage	3.3 V
Leakage current ($I_{leakage}$)	10 nA
Receive current (I_{rx})	11 mA
Sleep mode current (I_{sleep})	1.5 µA
Noise figure (NF)	**10 dB**

payload sizes and symbol configurations. The simulations are conducted over 50,000 iterations under an additive white Gaussian noise (AWGN) channel with varying initial conditions.

Figure 6.4 illustrates the SNR thresholds that determine the reliability of various SFs when using CR = 4/7. The results reveal that BER thresholds of 10^{-3}, 10^{-4}, and 10^{-5} serve as references for estimating SNR margins across different SF–CR combinations. The simulations use a carrier frequency of 868 MHz, multiple CR configurations based on Hamming codes, and a fixed bandwidth of 125 kHz to investigate long-range LoRa behavior.

We assume a processing clock frequency of 4 MHz for all computational operations. The path-loss exponent is set to $\alpha = 3$ to model urban environments and $\alpha = 4$ to model suburban areas. These assumptions enable us to capture realistic transmission behavior under varying spatial conditions and validate the energy model proposed earlier (Figure 6.5). The number of transmissions per ED is limited cyclically by duty cycle regulations. Typically, a maximum duty cycle of 1% is enforced. This constraint implies that after transmitting for a duration equal to the time-on-air (TOA), the ED must remain silent for 99% of the time. The communication delay includes both the idle period and the time required to receive acknowledgment. Therefore, larger packets result in longer TOA durations and consequently longer idle periods, reducing the number of possible transmissions before battery depletion.

To examine the influence of duty cycle on network lifetime, we analyze node behavior under duty cycle values of 1% and 0.1%. Figure 6.6 demonstrates the impact of payload size on the number of transmissions when using configuration (SF7, CR = 4/8). It can be observed that increasing data size reduces the total number of allowable transmissions for a battery capacity of 2,600 mAh. Hence, shorter packets are advantageous for prolonging the number of communication cycles. To predict lifetime under different energy constraints, we simulate nodes with battery capacities of 500, 2,600, and 3,500 mAh, each transmitting a fixed payload size of ten bytes per duty cycle. Figure 6.6 [8] shows that the node's lifespan is strongly affected by the

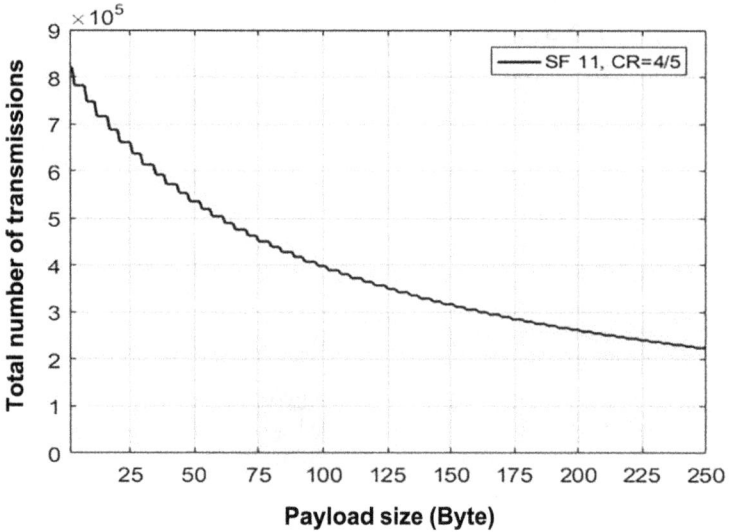

FIGURE 6.5   Total transmissions of LoRa EDs using (SF7, CR = 4/8).

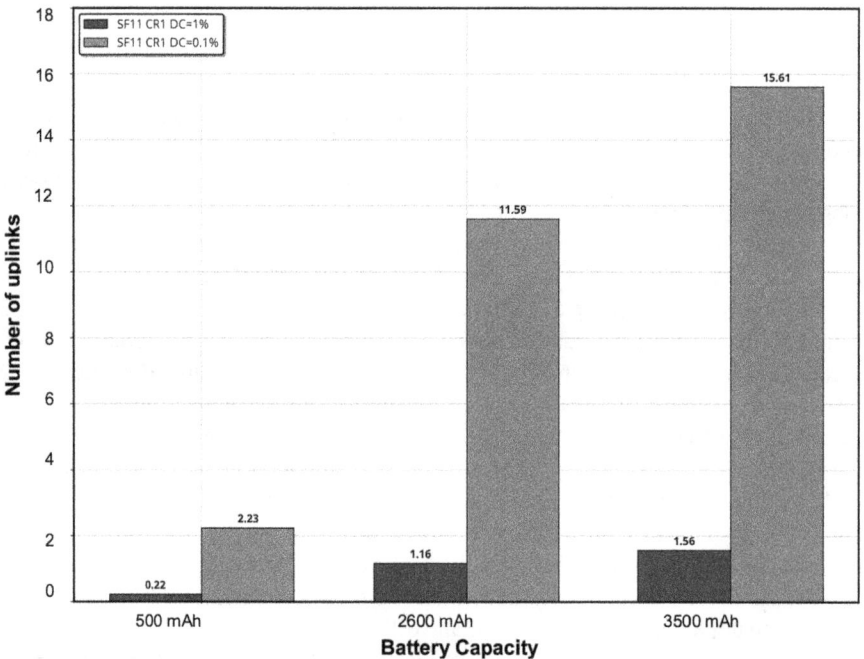

FIGURE 6.6   The lifetime of an ED using different battery capacities.

selected duty cycle. A reduced duty cycle (e.g., 0.1%) causes faster energy depletion as it allows more frequent transmissions, reducing the idle gap between successive transmissions. However, lifetime should be assessed not only in terms of time but also by the total number of successful transmissions.

Thus, energy-efficient design for LoRa EDs is a trade-off between communication frequency, packet size, and duty cycle compliance. A lower duty cycle extends the node's lifetime by reducing daily transmissions but also limits the rate of data delivery. Consequently, the lifetime of a LoRa ED can be interpreted as the result of balancing active and idle durations across regulatory and application constraints.

6.5 CONCLUSION

This chapter began by exploring the operational aspects of the communication chain in LoRa transmissions, with a particular emphasis on CSS modulation and channel encoding techniques. Building on this foundational understanding of the LoRa physical layer, we developed a comprehensive mathematical framework to model and estimate the energy consumption of LoRa-enabled end devices.

To validate the proposed communication and energy models, we conducted extensive MATLAB-based simulations across various network scenarios. These simulations encompassed both physical layer parameter variations and topological configurations typical of LPWAN deployments.

The resulting analysis offered valuable insights into the trade-offs between energy efficiency and transmission reliability, revealing critical operational thresholds under duty cycle constraints.

By specifying battery capacities and configuring different duty cycle levels, we were able to identify optimal transmission parameters—namely spreading factors (SFs) and coding rates (CRs)—that maximize node lifetime without compromising communication performance. These evaluations allowed us to estimate the total number of successful transmissions and the overall lifespan of a customized LoRa sensor node.

In summary, this work provides a reproducible energy evaluation methodology for LoRa-based communication, integrating analytical modeling and simulation to support energy-aware protocol design and deployment strategies in wide-area wireless sensor networks.

REFERENCES

1. I. Lamaakal et al. 'A deep learning-powered TinyML model for gesture-based air hand-writing simple arabic letters recognition'. In: *International Conference on Digital Technologies and Applications*. Benguerir, Morocco. Springer Nature Switzerland, 2024, pp. 32–42.
2. I. Lamaakal et al. 'A comprehensive survey on tiny machine learning for human behavior analysis'. In: *IEEE Internet of Things Journal* 12.16 (2025), pp. 32419–32443. DOi: 10.1109/JIOT.2025.3565688.
3. J. Haxhibeqiri. 'A survey of LoRaWAN for IoT: From technology to application'. In: *Sensors* 18.11 (2018), p. 3995. DOi: 10.3390/s18113995.
4. S. Kim, H. Lee & S. Jeon. 'An adaptive spreading factor selection scheme for a single channel LoRa modem'. In: *Sensors* 20.4 (2020), p. 1008. DOi: 10.3390/s20041008.
5. I. Lamaakal et al. 'Optimizing breast calcification detection in mammography using PyS-park: A big data and machine learning approach'. In: *International Conference on Data Analytics and Management*. London, UK. Springer, 2025, pp. 617–627.

6. M. Slabicki, G. Premsankar & M. Di Francesco. 'Adaptive configuration of LoRa networks for dense IoT deployments'. In: *IEEE/IFIP Network Operations and Management Symposium (NOMS)*. Taipei, Taiwan. IEEE, 2018, pp. 1–9. DOi: 10.1109/NOMS.2018.8406255.

7. I. Lamaakal et al. 'A TinyDL model for gesture-based air handwriting Arabic numbers and simple Arabic letters recognition'. In: *IEEE Access* 12 (2024), p. 76589–76605.

8. M. Bor & U. Roedig. 'LoRa transmission parameter selection'. In: *2017 13th International Conference on Distributed Computing in Sensor Systems (DCOSS)*. Ottawa, Canada. IEEE, 2018, pp. 27–34. DOi: 10.1109/DCOSS.2017.10.

9. B. Chaudhari & S. Borkar. *Design Considerations and Network Architectures for Low-Power Wide-Area Networks*. INC, 2020.

10. I. Lamaakal et al. 'Tiny language models for automation and control: Overview, potential applications, and future research directions'. In: *Sensors* 25.5 (2025), p. 1318.

11. C. F. Dias, E. R. De Lima & G. Fraidenraich. 'Bit error rate closed-form expressions for LoRa systems under Nakagami and rice fading channels'. In: *Sensors* 19.20 (2019), p. 4412. DOi: 10.3390/s19204412.

12. I. Lamaakal et al. 'A TinyML model for gesture-based air handwriting Arabic numbers recognition'. In: Edited by M. Mahmud, M. Lahby, M. Shamim Kaiser, P. Manzoni, and Y. Maleh. *Procedia Computer Science*. Vol. 236. Elsevier, 2024, pp. 589–596.

13. D. H. Kim, E. K. Lee & J. Kim. 'Experiencing LoRa network establishment on a smart energy campus testbed'. In: *Sustainability* 11.7 (2019), p. 1917. DOi: 10.3390/su11071917.

14. I. Lamaakal et al. 'Efficient gesture-based recognition of tifinagh characters in air handwriting with a TinyDL model'. In: *2024 Sixth International Conference on Intelligent Computing in Data Sciences (ICDS)*. Marrakech, Morocco. IEEE, 2024, pp. 1–8. DOi: 10. 1109/ICDS62089.2024.10756483.

15. I. Ez-Zazi, M. Arioua & A. El Oualkadi. 'Adaptive joint lossy source-channel coding for multihop IoT networks'. In: *Wireless Communications and Mobile Computing* 2020 (2020), p. 2127467. DOi: 10.1155/2020/2127467.

7 Security and Privacy in TinyML Applications

Mohammed R. Al-Matari and Qasem Abu Al-Haija

7.1 INTRODUCTION

The rapid growth of the Internet of Things (IoT), enabled by faster internet connectivity, has paved the path for integrating modern artificial intelligence (AI) techniques into edge devices, resulting in an urgent demand for smart autonomous edge appliances. This need has driven the rise of a new innovative field of machine learning (ML) called Tiny Machine Learning (TinyML), which extends the capabilities of ML to resource-constrained, ultra-low-power devices operating at the edge of the network, such as microcontrollers [1,2]. By allowing data processing, on-device analytics, and decision-making without the need for cloud computing, TinyML offers major benefits, including but not limited to reduced latency, more privacy, lower bandwidth consumption, and more energy-efficient usage.

TinyML has demonstrated transformative potential in various applications, including predictive maintenance in industrial IoT, wearables, smart farming, healthcare monitoring, and wildlife preservation [3,4]. All these use cases involve work via real-time inference and constant sensing in sensitive settings, highlighting the importance and complexity of security and privacy challenges. Unlike traditional ML that works in cloud computing, TinyML's working context has many constraints, such as limited resources, physical exposure, and low-trust environments, resulting in a vulnerable system with weak defense capabilities.

Despite the promising potential of TinyML, constrained computations of edge devices pose a barrier to traditional security and privacy-preserving mechanisms. Trivial operations on general-purpose CPUS, such as cryptography, are impractical with limited processing power with kilobytes of RAM. Similarly, privacy-enhancing approaches such as differential privacy and homomorphic encryption are infeasible in this case without any substantial modification. These limitations necessitate more directed, lightweight methods designed for the edge devices.

Moreover, AI models' deployment in safety-critical settings magnifies the possible risks. TinyML systems are prone to several attacks, including data poisoning attacks, model inversion attacks, adversarial input attacks, firmware tampering, and hardware-based side-channel exploitation. Consequences of such attacks range from privacy breaches to disastrous system failures, precisely in critical domains such as autonomous systems and healthcare [5]. Figure 7.1 shows the TinyML architecture layers and the corresponding threats at each layer. This highlights the need to address threats at each layer and consider threats at other surrounding layers. Therefore, the security and privacy of TinyML systems are emerging concerns that need to

DOI: 10.1201/9781003544449-7

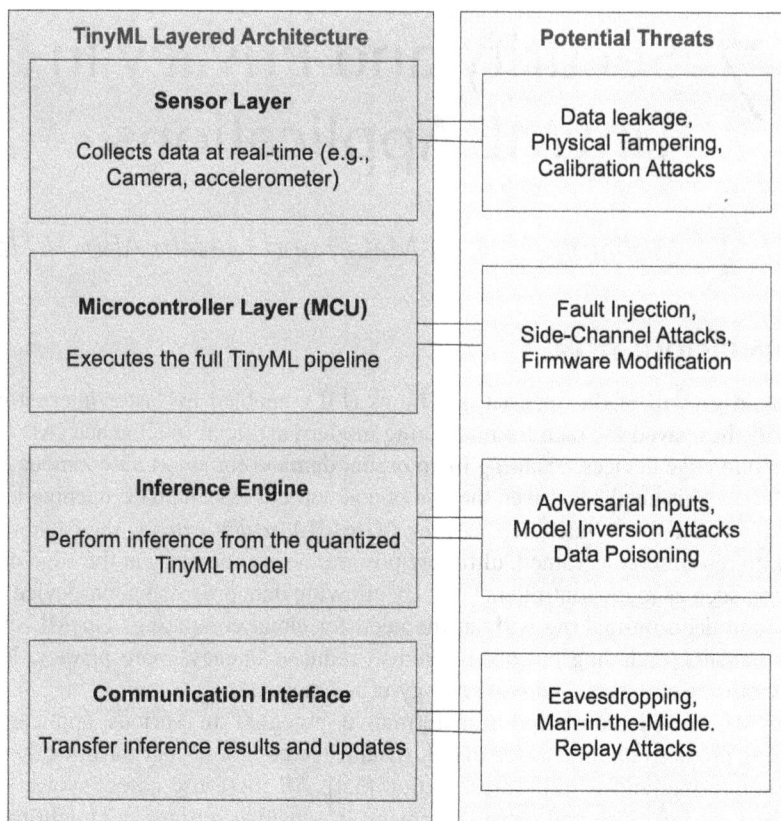

TinyML Layered Architecture	Potential Threats
Sensor Layer Collects data at real-time (e.g., Camera, accelerometer)	Data leakage, Physical Tampering, Calibration Attacks
Microcontroller Layer (MCU) Executes the full TinyML pipeline	Fault Injection, Side-Channel Attacks, Firmware Modification
Inference Engine Perform inference from the quantized TinyML model	Adversarial Inputs, Model Inversion Attacks Data Poisoning
Communication Interface Transfer inference results and updates	Eavesdropping, Man-in-the-Middle. Replay Attacks

FIGURE 7.1   TinyML layered architecture and potential threat at each layer.

be holistically addressed throughout their phases—from data collecting and model training to the deployment phase and, finally, lifecycle management.

This chapter offers a comprehensive, thorough overview of TinyML applications' security and privacy issues, catering to industry practitioners and academic researchers. The chapter aims to provide the reader with a thorough overview of the challenges and promising solutions in the field by covering the existing threat landscape, privacy-preserving ML techniques, cryptographic constraints, and secure deployment frameworks. Usually, TinyML systems' privacy is discussed in isolation, technically, and not in a holistic approach in existing literature. However, this chapter aims to address TinyML security and privacy as an integrated threat landscape. This chapter bridges edge AI, privacy-enhancing technologies, and embedded systems altogether. It surveys state-of-the-art vulnerabilities and mitigations, provides overviews of structured frameworks, provides design guidelines, and exhibits practical use cases to guide future development. Its key contribution equips researchers, engineers, and policymakers with a practical roadmap toward the trustworthy and secure deployment of TinyML at scale.

7.1.1   WHY THIS CHAPTER?

This chapter is motivated by the pressing need to:

- Understand the rising security and privacy concerns from the deployment and architecture of TinyML.
- Introduce and evaluate the lightweight cryptography protocols suitable for edge devices.
- Present current privacy-preserving mechanisms such as differential privacy, federated learning, and adaptation to embedded systems.
- Illustrate real-world case studies and their security incidents, mitigation, and best practices.
- Highlight the potential of cross-disciplinary collaboration across AI, embedded systems, and cryptography communities.

7.1.2   CHAPTER STRUCTURE

The structure of this chapter is organized as follows:

- Section 7.2 Background and Fundamentals: provides the reader with the fundamentals and background of TinyML system architecture, development lifecycle, security challenges, and threat model.
- Section 7.3 Security Threats and Attack Surfaces in TinyML Systems delves into the security threats on TinyML from various sides, such as hardware-level, software-level, network-level, and ML-specific threats.
- Section 7.4 Privacy Concerns in TinyML Applications reviews the privacy concerns in TinyML applications.
- Section 7.5 Lightweight Cryptography and Secure Protocols for TinyML outlines the lightweight cryptographic standards and protocols used for embedded systems.
- Section 7.6 Privacy-Preserving and Secure ML Techniques for TinyML explores the privacy-preserving techniques for secure ML and TinyML, such as homomorphic encryption, federated learning, and differential privacy.
- Section 7.7 Secure TinyML Frameworks, Toolchains, and Case Studies surveys secure TinyML frameworks and toolchains and presents real-world case studies.
- Section 7.8 Design Guidelines and Best Practices provides practical design guidelines and best practices for a secure TinyML and privacy-aware development.
- Section 7.9 Research Challenges and Future Directions highlights the emerging open research directions, future challenges, and developing standards in the domain.
- Section 7.10 Conclusion concludes and summarizes the key insights and advice on building a trustworthy TinyML from the ground up.

7.2   BACKGROUND AND FUNDAMENTALS

7.2.1   TINYML ARCHITECTURES AND DEPLOYMENT CONTEXT

TinyML is a machine learning model that operates on resource-constrained edge devices, particularly edge sensors and small-scale IoT devices with microcontroller units (MCUs). A typical TinyML system has one or more data sources, mainly sensors (e.g., camera and microphone) passing data into an on-board processing unit (MCU) in which the TinyML model is run. The MCU runs the inference engine for a pre-trained TinyML model, then an output inference produces the results, either actuating it or communicating the results (Figure 7.1). Unlike smartphones, MCUs that run TinyML models have no operating system (usually a lightweight real-time operating system, RTOS), which operates in a limited memory range of tens or hundreds of kilobytes. For instance, several well-known TinyML platforms utilize Arm Cortex-M CPUs, with an SRAM < 256 KB and a low clock speed. Thus, efficient model architecture and software are required to work in such severe conditions [6].

Despite the size, TinyML provides great advantages to edge devices, enabling real-time, on-site AI inferences without continuous cloud connectivity [6]. This is critical for industrial sensors, health wearable monitors, and wake-word detection such as Amazon Alexa, Apple Siri, or Google Assistant. In those cases, streaming data back to the cloud for inference is infeasible due to bandwidth, response latency, and privacy concerns. TinyML is designed to operate on edge devices under low-power consumption settings. For instance, a simple keyword spotting TinyML model can be deployed on a coin-cell-powered Arduino system continuously listening for vocal commands [7]. Typically, such architecture utilizes optimized libraries (e.g., Arm CMSIS-NN, TensorFlow Lite). Additionally, they could incorporate matrix operations hardware accelerators. However, these platforms lack the large edge devices, such as virtual memory or process isolation. In summary, TinyML architectures have enabled the distribution of AI to edge devices and act as the basis of an "Edge AI" paradigm in which AI is transferred from the large cloud data centers into edge devices into embedded environments by bringing the complete cycle (sensing, computation, and actuation) into a single edge device.

7.2.2   MACHINE LEARNING LIFECYCLE ON EDGE DEVICES

TinyML application development consists of several phases: data acquisition and deploying the trained model into the environment. A typical TinyML lifecycle involves:

1. **Data Acquisition:** Collecting data from the target environment, such as images from a camera or audio clips from a microphone, and then annotating the data for the training phase. Because TinyML performs specific tasks (e.g., detecting a sound), the data quality plays a crucial role in the model's performance.
2. **Preprocessing and Feature Extraction:** Preprocessing the sensors' raw data into a suitable feature for the ML model. For instance, sound detection applications include signal processing, noise reduction, and dimensionality reduction techniques. Preprocessing can occur on-device, such as

simple filtering or normalization, before feeding the signal into TinyML for inference.

3. **Model Training:** The collected data is used to train the model (a compact model consisting of a small neural network, NN, for instance) using a standard framework on the cloud or a PC because edge devices are too resource-limited to perform on-device training. Thus, the training is performed in development environments with no computing resource limitations. The developed model architecture is designed to fit the limitations of the MCU (e.g., using shallow neural networks or tiny ConvNets).

4. **Model Optimization:** Shrinking the model's size to load it onto the MCU and run it. Common techniques include pruning (dropping unnecessary weights), model distillation or compression, and quantization (reducing the model precision, e.g., from 32-bit to 8-bit integers). Optimizations trade between a small drop in accuracy and fitting the model into the MCU memory. For example, quantizing the model into 8-bit integers could save space so that the model runs on the MCU memory while still having high accuracy [8].

5. **Deployment Format Conversion:** In this phase, the model is converted into a suitable MCU format, such as flatbuffer or C array (for TensorFlow Lite Micro). Then, the model is compiled and linked to the edge device firmware.

6. **Flashing and Deployment:** The model weights and code are flashed into the MCU flash (often 512 KB-1 MB). This way, the model and inference runtime are programmed into the microcontroller. Sensor inference is initialized on boot, and the ML is moved to memory.

7. **On-Device Inference and Monitoring:** The model runs on the device at this stage, and inference occurs locally on the sensor data coming in real-time. Some TinyML deployments include monitoring feedback: the model's performance is logged to identify further model updates and improvements, if needed. However, the TinyML model update process on deployed devices is challenging, as it may require an over-the-air (OTA) firmware update if supported by the MCU (Figure 7.2).

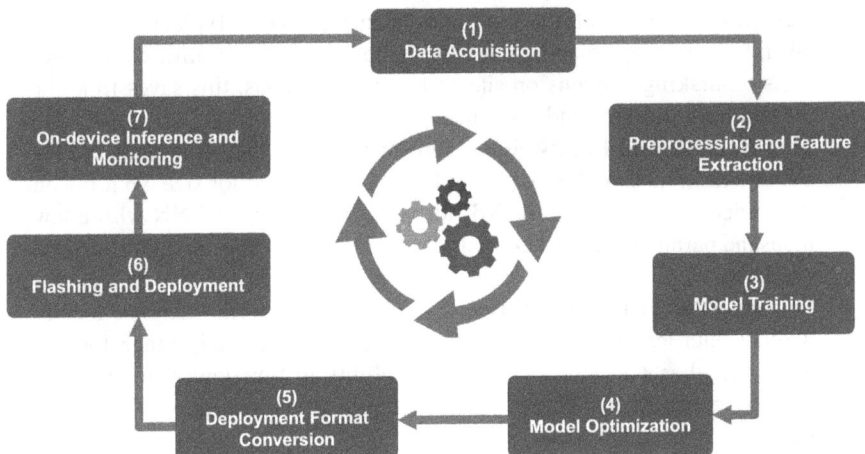

FIGURE 7.2 TinyML lifecycle phases.

During the TinyML lifecycle, engineers use specialized tools to streamline the development process. For instance, Google TensorFlow Lite Micro enables model testing on-device. Other platforms, such as Edge Impulse or Qeexo AutoML, provide an end-to-end workflow for data collection, training, and deployment. A critical aspect of the lifecycle is the train-deploy gap. As the training happens on an off-device, the deployment onto the device requires careful simulation of the MCU's numerical behavior and storage limitations, precisely when applying optimization techniques to ensure that the model will work properly under the MCU constraints. The TinyML lifecycle is iterative; engineers must go back to collect data or redesign the model if the accuracy is unsatisfactory. In summary, standard ML and TinyML lifecycles align in several ways, except that the TinyML process has additional steps of model compression into edge devices and validation under constrained resources in real-time environments.

7.2.3   KEY CHARACTERISTICS OF TINYML

Several characteristics distinguish TinyML systems from conventional ML or cloud-deployed ML. TinyML characteristics are as follows:

- **Limited Memory and Compute:** TinyML is deployed on MCUs equipped with kilobytes of RAM, flash, and often a tens of MHz CPU speed. Therefore, models must be small to fit on the limited resources. For instance, a smartphone has 10^3 times more memory than an MCU [9].
- **Ultra-Low Power Operation:** MCUs are designed to run on low power, usually milliwatt or microwatt, so TinyML has been extremely energy-efficient. As TinyML systems may always involve on-inference (e.g., always listening for a keyword), they need to save power by any means, such as saving the energy cost of data streaming over wireless connections. For example, it has been illustrated that a coin-cell battery can power a keyword spotting system for months running on Cortex-M4 [7].
- **Real-Time Settings:** TinyML applications are mostly real-time; therefore, they must respond with the minimum latency possible. Typically, the time taken from sensing to actuation is in the range of tens of milliseconds. As decision-making happens on-site and is instantaneous, this saves time for the end-user, unlike cloud inference.
- **Task-Specific and Lightweight Models:** Since TinyML is specialized for a single task, its complexity is kept low and optimized for size, so it fits on the device. For instance, TinyML for voice would have a CNN with a few thousand parameters, unlike the cloud version, which may be multi-purpose and large.
- **Privacy-Preserving and Offline:** The ability to perform inference locally without data leaving the device is an inherent privacy advantage for end users. Another TinyML attribute is the ability to function offline without needing connectivity. Thus, systems are robust and operate with no connectivity to the cloud services.

In summary, the characteristics of TinyML revolve around task specificity, continuous real-time operating, and limited resource usage. With the billions of microcontrollers around (250 billion by a study conducted in 2022), and this number is continually increasing, TinyML traits enable the edge devices to incorporate ML into their operations. With such benefits offered by TinyML, it is still accompanied by significant challenges, such as privacy and security, as discussed next.

7.2.4   SECURITY CHALLENGES IN EDGE AI VS. CLOUD AI

Deploying TinyML provides a distinct security advantage in which data is processed locally on-device, unlike in cloud AI, where data is transferred from users or sensors to the central server for training or inference. Still, TinyML has several privacy vulnerabilities. Table 7.1 highlights the key differences in security considerations between Edge AI (TinyML and distributed) and Cloud AI (centralized).

The differences in the security issues between TinyML and cloud-based ML require designers to shift their mindset into a more careful approach to security techniques used in TinyML, such as sensor data authentication, encrypted storage, etc. However, the implementation of such techniques on low-resource hardware is a challenge. Next, threat models to TinyML are outlined for consideration during TinyML systems analysis.

7.2.5   THREAT MODELS RELEVANT TO TINYML

Given the operating environment of embedded systems, threat modeling would vary depending on the malicious attackers' capabilities. Typically, threat models range from hardware and physical attack surface to the software side and ML-related attacks. Threats are broadly categorized into the following three classes:

1. **Hardware and Physical Attacks:** These attacks build on the assumption that the attacker has physical access to the device or a proximity to it. This is a realistic scenario since edge devices mostly work in an outdoor environment. Threats to the hardware include device tampering, fault injection, side-channel attacks, and eavesdropping on output/input attacks.
2. **Software and Firmware Attacks:** Attacks targeting the MCU software usually aim at the device firmware or ML model code. Such attacks can be performed remotely without the need of the attacker's physical presence nearby the device. Such attacks include firmware or unauthorized model updates, memory corruption, and code execution.
3. **Network and Communication Attacks:** Attacks on communication channels exploit vulnerabilities in wireless communications protocols and data transmission. As edge devices utilize low-power communication protocols such as LoRa, Zigbee, etc., they are prone to manipulation, unauthorized access, and interception by attackers without physical proximity required.
4. **ML-Specific Attacks:** Such attacks are aimed at the ML itself by leveraging its properties and the data used. Such attacks include data poisoning, model extraction, model misuse or repurposing, and adversarial examples.

TABLE 7.1

Cloud AI vs Edge (TinyML) AI Security Considerations Comparison

Aspect	Cloud AI	Edge AI
Data transmission	Risk of data sent off-device and central storage breaches	Data is kept on-device and provides better privacy. Local breach impacts affect device data only.
Physical security	Central datacenters and servers are highly protected; users can't access cloud hardware physically	Edge devices mostly operate in an open environment; they are at attackers' reach. They can be physically tampered, accessed, or damaged.
Computing resources	No computation limitations: thus, protection techniques can be implemented, such as strong encryption, anomaly detection, etc.	Limited resources—it is infeasible to implement heavy-computing security features, only lightweight techniques can be implemented.
Attack surface	Cloud has fewer access points, such as APIs/cloud interfaces. Still, they are attractive, as one breach can expose massive data	Exposed in both ways; physical interfaces and via the network (in case connected). However, small targets are hard to manage.
Update mechanism	Cloud updates are instantly sent to all users at once—centralized updates	Distributed updates; each MCU must be updated if capable, resulting in some MCUs running outdated firmware
Adversarial ML	Physical sensors are not easily affected; however, adversaries can remotely send corrupt data to the cloud model. An input monitoring can be applied in such a case.	Sensors are directly influenced by either noise or light. Additionally special crafted physical adversarial examples can be crafted, and these ones are hard to detect.
Intellectual property	Model parameters are hard to steal as they are stored on secure servers. Inverting training data is challenging without breaching the server.	The model is within the device flash memory and can be extracted via firmware extraction or hardware attacks. They are more vulnerable to theft [7].

In the threat model definition, one can assume the worst-case scenarios in which the attacker has full capability. Another assumption is that attackers can gain physical access to the device, as it is widely distributed and placed outdoors. Thus, attackers gaining access to devices enables them to perform analysis on the device, test their code, and observe the device behavior under attacks to improve their attacks. For TinyML, engineers and researchers need to prioritize the various attacks based on their severity using methodologies such as CVSS (Common Vulnerability Scoring System) to know which threat to mitigate first. All in all, TinyML combines the usage of ML and IoT fields, thus inheriting threats from these two fields. Additionally, constrained by the hardware of those edge devices. A balanced security model guides engineers in developing a defense mechanism to achieve sensible security for TinyML systems under their working conditions.

7.3 SECURITY THREATS AND ATTACK SURFACES IN TinyML SYSTEMS

TinyML systems are inherently exposed to a wide range of vulnerabilities due to their decentralized operation nature. The deployment conditions, low memory, default no-encryption settings, and operating system absence all enlarge the attack surface on the systems over their multiple working layers. The section explores the possible attack vectors along with their threat types in their architectural layer: hardware layer, software/firmware layer, communication layer, and machine learning model layer.

7.3.1 OVERVIEW OF ATTACK VECTORS

Unlike the cloud AI systems, TinyML works in an exposed area, making it prone to attacks of different types, including hardware interface exploitation, insecure communication channels (e.g., update channel), or the model inference behavior. TinyML has been increasingly targeted by malicious adversaries due to its accessibility and working conditions [10]. Furthermore, implementation of security measures such as full-stack, sandboxing, and runtime anomaly detection is infeasible due to the constrained nature of MCUs (e.g., limited computing power and memory) [11]. Figure 7.1 shows the four different levels of threat/attack types under each category, as shall be explained in the coming text.

7.3.2 HARDWARE-LEVEL THREATS

Hardware-level threats aim to leverage vulnerabilities in the hardware layer, which include the following:

a. **Device Tampering:** The attacker can gain access to the data or model stored in the device via the debug ports (SWD/JTAG) if not locked. They can physically modify the device and get read/write access to the firmware. Lack of secure boots and readout protection enables attackers to close the models easily. A noticeable widespread misconfiguration of the interfaces in IoT devices with TinyML systems shows a similar trend [12].

b. **Side-Channel Attacks:** Attackers can infer data from the MCUs by monitoring their operations, such as timing or electromagnetic emissions. Usually, side-channel attacks aim to attack cryptographic keys; however, recent studies showed that it is possible to extract the ML model in a similar fashion [7]. A study by Batin et al. illustrated the feasibility of extracting neural network (NN) architecture via electromagnetics SCA on a TinyML MCU Cortex-M3 chip [13]. Additionally, model confidentiality can be compromised, uncovering its inputs and weights via unprotected power traces [14].

c. **Fault Injection:** Attacks can control the MCU to misbehave in a certain way by inducing glitches such as clock spikes or laser flashes. Such misbehaviors can be security check skipping or model prediction deterioration at critical times. Bit-Flip attacks can degrade the model performance significantly by corrupting quantized model critical weights; such precision-targeted attacks are catastrophic in critical applications such as health monitors or smart locks [15]. Unfortunately, countermeasures such as redundant computation or secure fault detection don't exist on TinyML systems hardware, mainly due to cost constraints.

7.3.3 SOFTWARE-LEVEL THREATS

Software attacks mainly revolve around the update and memory limitations of the TinyML systems. Such attacks include the following:

a. **Unauthorized Model or Firmware Update:** An attacker can attempt to load corrupted firmware or model onto the device, considering that the TinyML device supports remote update and has remote updates mechanisms such as UART or OTA. For instance, a Trojaned model can be injected to leak sensitive data or misclassify specific inputs to the system. It has been shown that basic CRC checks are not enough to defend against such a threat, and embedded environments lack any cryptographic signing.
b. **Memory Corruption and Code Execution:** As TinyML has no operating system (i.e., no protection provided), applications run their code written mostly in C/C++, which can lead to issues such as buffer overflow or firmware memory vulnerability. Attackers can exploit that to inject malicious code for adversarial purposes, such as disabling the model, or worse, to repurpose the model for harmful activity. Such threats are damaging due to the lack of OS-level privilege isolation and memory protection (Figure 7.3).

7.3.4 NETWORK-LEVEL THREATS

Network-level attacks exploit the communication channels between the edge device and central servers precisely during update times. As edge IoT devices use lightweight transport protocols, making them weak against attacks.

a. **Eavesdropping on Outputs/Inputs:** Gaining access to the communication interface (e.g., wireless transmission or SPI bus) enables attacks to acquire sensitive information. For example, when someone is at home, Eavesdropper can learn when the TinyML sends an alert over the BLE. Thus, considering physical layer security techniques is crucial, such as encryption or pairing.
b. **Replay Attacks:** TinyML systems transfer data mostly using low-power wireless connectivity such as LoRa, Zigbee, or BLE. Encryption over such channels is weak if not omitted, which enables malicious attackers to intercept data to use later for replay attacks in which they send legitimate requests to trick the system. LoRaWan replay attacks have been demonstrated to bypass lock door defense using the earlier captured authorization packets [16].

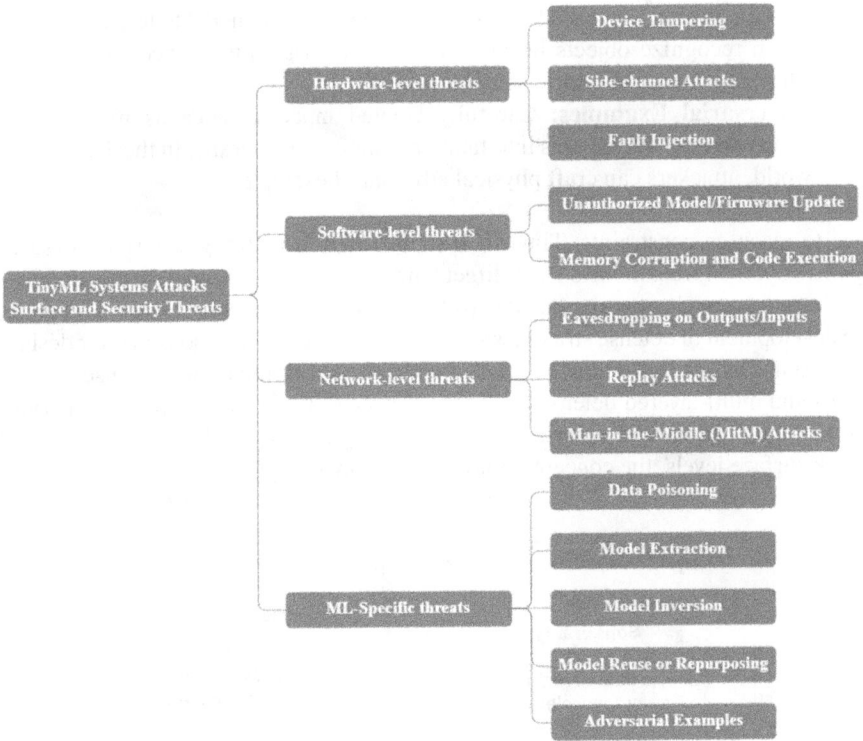

FIGURE 7.3   TinyML attack surface levels.

c. **Man-in-the-Middle Attacks:** Lack of shared authentication leaves space to attackers to inject faulty values or manipulate the model parameters during OTA updates. It has been demonstrated that MitM attacks on MQTT-based edge nodes can trick the models into unauthorized behavior [17].

7.3.5   ML-SPECIFIC THREATS

a. **Data Poisoning:** An attacker can inject corrupted data into the model that is used in the periodic retraining or calibration (say under federated learning settings). An attacker can contaminate the training data at the collection stage by mislabeling examples for instance.

b. **Model Extraction:** Model extraction is critical precisely when that model is trained on expensive data, which is considered intellectual property. This can be done by query attacks (via providing input and monitoring the output) or via side-channels attacks (as mentioned).

c. **Model Inversion:** Model inversion is the reconstruction of model inputs from observed outputs. Such attacks have successfully extracted training data from quantized models through activation patterns or output logits [18].

d. **Model Misuse or Repurposing:** An attacker can use the model to behave out of its intended purposes. For example, in a camera-based TinyML system, an attacker could feed custom images into the model to test whether it can recognize objects beyond those it was originally trained to detect. Thus, attackers use it for spying purposes [18].

e. **Adversarial Examples:** Carefully created inputs can act as malicious inputs driving ML models into faulty predictions. Generally, in the TinyML world, attackers can craft physical adversarial examples.

Due to resource constraints, TinyML systems often lack strong security measures against layered threat models from direct hardware attacks, software/firmware pipeline insecurity, and network-related gaps to finally model-specific threats. Therefore, the development of defense strategies should consider such limitations in their design. Understanding the broad attack surface is crucial to building robust defense mechanisms and multi-layered defenses, considering verifying firmware updates, encrypting communication, and securing bootloaders. Figure 7.4 concludes the section on attack surface levels, the conceptual layers of a TinyML system.

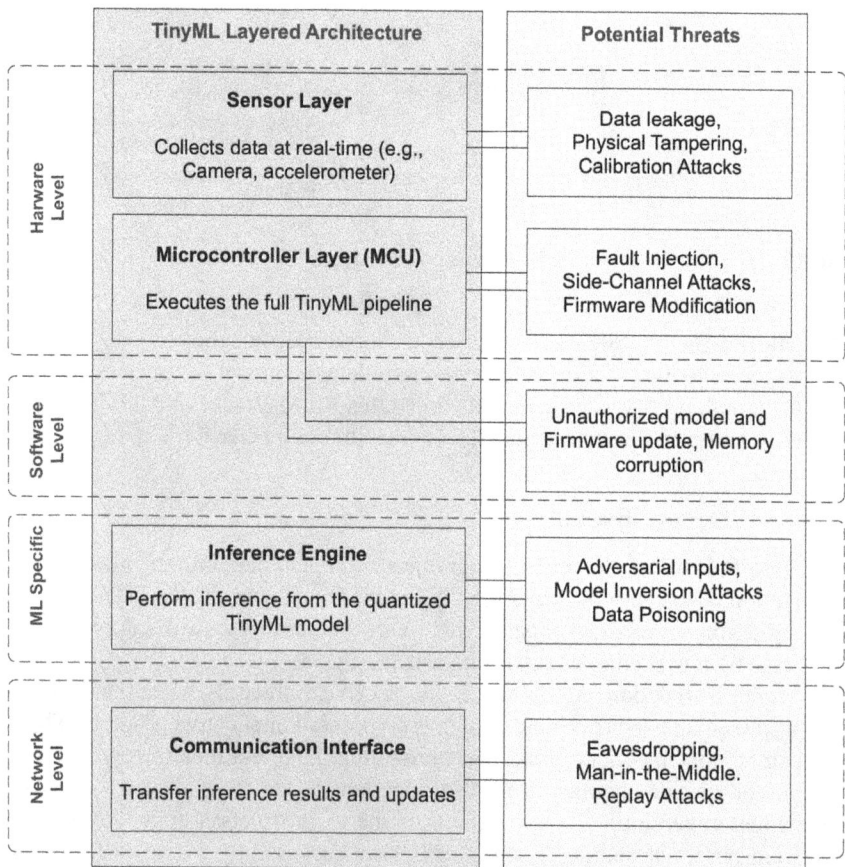

FIGURE 7.4   TinyML conceptual layers along their threat levels.

7.4 PRIVACY CONCERNS IN TinyML APPLICATIONS

With the migration of traditional ML from cloud to edge devices, they offer an opportunity for enhanced data privacy. TinyML systems perform tasks on-device; thus, no data is transferred to the cloud, potentially keeping user privacy. However, it has its own distinct challenges, such as user consent management, data retention, and emerging threats posed by adversarial machine learning. This section delves into the privacy implications of TinyML, covering real-world scenarios and use cases, practical mitigation techniques, edge-based inference, and key variations between TinyML and cloud-based ML systems.

7.4.1 PRIVACY IMPLICATIONS OF ON-DEVICE LEARNING AND INFERENCE

TinyML systems perform inference on the device without the need to transfer the data back to a central cloud server, as in the case of cloud-based AI systems. This advantage of TinyML over cloud-based ones eliminates the possibility of any threat to the network, such as packet sniffing, interception, or central data breaches on the cloud [19]. However, such an advantage heavily depends on the safeguards of devices. Lack of storage encryption or secure boots can hinder the privacy gained. With attackers having physical access to the MCU, they can easily extract stored data in the MCU [20,21]. Additionally, as TinyML is trained off-line, engineers prioritize inference privacy in this case. However, during training, data is collected, sabotaging privacy unless federated learning is used, in which the end user stores their data locally and still contributes to the model training, or using continual learning. Even then, model inversion (i.e., inference results) can still leak data and sensitive information in the absence of strong safeguards, particularly when using unsecured APIs or bluetooth low energy (BLE) connections.

7.4.2 CASE STUDIES: PRIVACY IN REAL-WORLD TinyML APPLICATIONS

TinyML helps maintain privacy in several real-world cases and applications ranging from the healthcare field to security and smart homes.

1. **Surveillance and Smart Cameras:** TinyML models are deployed in several surveillance application systems for face recognition, motion detection, and abnormal movement classification [21]. For example, surveillance security cameras that send an alert when detecting a suspicious movement (e.g., "Human detected") instead of the raw video data, thus preserving privacy. Such systems minimize privacy accuracy and reduce latency by local data processing.
2. **Health Monitoring Wearables:** TinyML is deployed on several health wearables such as smart watches and fitness bands. They analyze the user's health information, such as ECG signals and blood oxygen value. Privacy is maintained as the data is only present on the device unless shared explicitly [22]. For instance, companies have integrated local fall detection and heartbeat irregularity models into their health wearables, aligning with privacy preservation principles.

3. **Smart Homes and Voice Interfaces:** Smart Homes with voice assistants use TinyML for local wake-word detection, such as "Hey Google," which omits the need for audio recording and data transmission over the network. This addresses the everlasting debate on privacy in voice assistants [23].

TinyML preserves privacy in those cases; however, it has been demonstrated that with weak, unprotected edge models, attackers can leverage reverse engineering to expose detection thresholds and, worse, exploit output to learn user behavior patterns and routines on their camera [24]. Researchers found that using adversarial audio patterns can activate voice assistants and fool them into misbehavior, thus proving that even local inference may not always be enough to prevent misuse. Users' awareness and control over these devices and what is permitted to be shared are critical to keep their privacy.

7.4.3 DIFFERENTIAL PRIVACY AND FEDERATED LEARNING IN TINYML

In order to hide the details of examples used in the training process, differential privacy (DP) is an assurance that the output doesn't disclose information. However, implementing this technique on MCUs is infeasible from a computational cost perspective, as it is an additional overhead to introduce noise to the output and clip it, particularly on low-resource edge devices [25]. A feasible approach, such as ε-local DP, which introduces an integer noise, is a promising exploration and is at an early phase, precisely in the area of sensor-based systems. Another approach is Federated Learning (FL), which is a suitable approach to the nature of the TinyML environment. In FL, multiple devices cooperate to train a global model on their local data and exchange only model parameters with the aggregation central server, as shown in Figure 7.5 [26]. Therefore, data doesn't leave the device, and privacy is maintained. It has been shown that a combination of secure aggregation and differential privacy can effectively maintain user privacy even across millions of participants' devices.

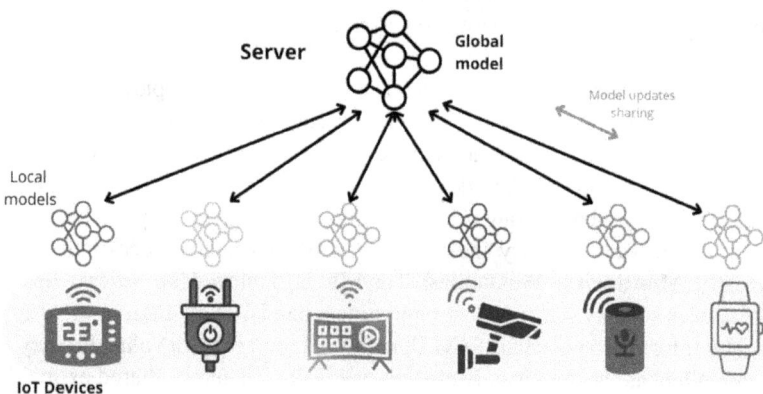

FIGURE 7.5 Federated learning model in an IoT environment for TinyML [26].

Adaptations of traditional FL to TinyML must overcome communication and energy challenges. Projects such as Flight [27] enhance efficiency via heuristic-based participants, paving the way for practical federated learning on low-power edge devices such as Cortex-M. Nevertheless, FL is not fully immune to privacy attacks; it is prone to gradient inversion attacks that enable attackers to extract training data examples through the model's shared updates [28]. Additionally, malicious participants or devices may contribute to poisoning the updates via a data poisoning attack or perform a model inversion attack since they gain access to the global model updates [29].

7.4.4   DATA RETENTION, ANONYMIZATION, AND USER CONSENT

Unlike the cloud-based AI systems, TinyML systems have no standard or clear protocol on user data collection and storage. While many devices would only store data temporarily within their buffers to discard later after that. Absence of a user interface limits users' ability to control device data-related behaviors. For instance, a wearable healthcare device may store user motion data for days or even weeks for trend analysis without prior user consent.

Another challenge is anonymization. Due to the rapid data acquisition and discarding in TinyML, it is seldom infeasible to implement anonymization techniques such as k-anonymity or I-diversity. Nonetheless, a feasible approach is data minimization—i.e., only collecting data that is essential for accurate inference only. Most embedded systems are limited, if not lacking, in screens to interact with the users. Thus, users can have the ability to learn about their privacy terms or even accept them. A proposed solution is to enable physical inferences (e.g., buttons) or pairing with smartphones to interact with the TinyML device [30]. It is still a gray area, and it is very challenging to achieve meaningful informed consent from users with limited interactions.

7.4.5   COMPARING TINYML AND CLOUD AI PRIVACY

TinyML inherently improves user privacy through local data processing. However, it comes with trade-offs. Could AI systems provide benefits of central governance, mature access control, and compliance with global standards such as the Health Insurance Portability and Accountability Act (HIPAA) [31] and the General Data Protection Regulation (GDPR) [32]? Nevertheless, centralized systems expose data during transmission and are vulnerable to breaches and insider threats [33]. On the other hand, TinyML escapes central data aggregation but lacks three main privacy measures: formal privacy protection, secure key management, and log auditing. An additional threat emerges when the edge device is compromised; localized data are susceptible to extraction. Figure 7.6 shows a comparison between TinyML and Cloud AI.

Feature	TinyML (local AI)	Cloud AI
Data Locality	Data is stored on-device	Data is stored in centralized datacenters
Internet Connection	Not required for inference	Required for inference
Consent Mechanism	It lacks consent opt-in/out user	It includes explicit opt-in/out mechanisms via forms or portals
Encrypted Inferences	Partial; varies based on device support	Fully and Strong; through TLS/SSL
Training Data Secured	Vulnerable	Secured
Standard Privacy Compliance	Immature	Strong

FIGURE 7.6   TinyML vs. cloud-AI privacy features comparison.

7.5   LIGHTWEIGHT CRYPTOGRAPHY AND SECURE PROTOCOLS FOR TinyML

With the growth in usage of TinyML systems in critical and sensitive environments such as healthcare, it is essential to ensure security using cryptography-based security mechanisms. However, conventional cryptographic protocols are infeasible in the harsh conditions of TinyML: limited RAM, no hardware acceleration, low power, and real-time requirements. This section outlines the lightweight cryptographic protocols, their challenges, related international standards, and efficient protocols for secure TinyML systems.

7.5.1   Cryptographic Constraints in Tiny Devices

Typically, TinyML systems run on microcontroller units (MCUs) like RISC-V or Arm Cortex-M cores, which are limited in computation power—often their RAM and flash memory are in the range of tens to hundreds of kilobytes. Such devices have no standard cryptographic hardware acceleration or instruction sets; thus, operating standard algorithms (e.g., AES-256, RSA2048, or SHA-512) on these MCUs is infeasible as they demand high computation and memory. For instance, a standard AES-128 requires 3 KB of ROM and needs >200 cycles per byte for encryption on the basic MCUs [34]. Moreover, efficiency is another crucial constraint. Running cryptographic operations on a battery-powered device can deplete the battery and shorten its lifespan. Thus, to maintain compatibility and usability, there is a need for lightweight cryptography that optimizes security, resource usage, and performance [35].

7.5.2   Lightweight Cryptography Standards: NIST and ISO

With the emerging demand for efficient cryptography in limited-resource edge devices, standardization bodies have announced dedicated standards:

- **NIST Lightweight Cryptography (LWC) Project (2018–2023):** a competition for lightweight authenticated encryption evaluation for Authenticated Encryption with Associated Data (AEAD) for low-resource TinyML devices. In 2023, the final standard was selected (ASCON) for its balance between security, speed, compactness, and performance [36].
- **ISO/IEC** 29192: An international standard that outlines many lightweight cryptographic algorithms for hash functions, MACs, asymmetric encryption, block ciphers, and stream ciphers. It involves algorithms such as CLEFIA, PRESENT, and hash functions designed for MCUs [37].

These guidelines provide developers with several options to select appropriate ciphers that meet compliance with standards and efficient for TinyML systems.

7.5.3   SYMMETRIC ENCRYPTION: SPECK, SIMON, AND ASCON

Low complexity and high-speed symmetric encryption are still the preferred approach for TinyML. The best-known candidates are

- **SPECK and SIMON:** These are lightweight block ciphers developed by the USA NSA in 2013 for both hardware (SIMON) and software (SPECK). SPECK is highly efficient working on 8-bit MCUs and ~65 cycles/byte [38]. Nonetheless, many governmental institutions and open-source projects avoid using them due to transparency and cryptographic trust issues.
- **ASCON:** ASCON combines both integrity and confidentiality by authenticated encryption provision. It has been introduced as the NIST LWC standard in 2023.
- A comparison of the four different cryptographic algorithms is illustrated in Table 7.2, focusing on their code size requirement and encryption efficiency (cycle/bytes).

7.5.4   LIGHTWEIGHT AUTHENTICATION: MACs AND HASH-BASED SIGNATURES

TinyML systems rely on the following authentication techniques for message integrity and authentication:

- **Message Authentication Codes (MACs):** Hash-based MACs (HMAC) and SHA-256 are commonly used; alternative resource-efficient options are the LightMAC and Chaskey for MCUs [39]. These MACs have a significantly smaller code size (<1.5 KB) and memory usage.
- **Hash-Based Signatures:** Post-quantum secure, stateless options such as SPHINCS+ are still under research and not yet mature for practical deployment.

For asymmetric cryptography, curves like Curve25519 and Ed25519 offer better advantages like small key size and faster operations, though they are computationally heavy on ultra-low power devices [40].

TABLE 7.2

Lightweight Cryptographic Algorithms Comparison

Cipher	Type	Encryption Speed (Cycles/Byte)	Code Size (Bytes)	Security Notes
ASCON	Authenticated AEAD	~110	~2,500	NIST LWC winner (2023); resistant to differential/ linear cryptanalysis
SPECK	Block Cipher	~65	~1,500–2,000	Highly efficient; not widely used due to trust/ transparency concerns (NSA origin); discontinued in IETF
SIMON	Block Cipher	~75 (hardware-optimized)	~2,000–2,500	Hardware-optimized; not widely used due to trust/ transparency concerns (NSA origin)
AES-128	Block Cipher	~240 (Cortex-M3, no HW accel)	~3,000–5,000	Robust security; computationally extensive for microcontrollers without AES-NI

7.5.5 SECURE BOOT AND FIRMWARE UPDATE MECHANISMS

Firmware and model updates in TinyML devices are sent over-the-air (OTA) and exposed to threats unless they are secure and validated cryptographically.

- **Firmware Authentication and Integrity:** The structured firmware meta-data, along with cryptographic hashes and public-key verification, can be performed using IETF SUIT and NIST SP 800-193. Arm's TF-M or STM32Cube is a lightweight implementation that integrates prior techniques for secure updates [41].
- **Secure Boot:** It guarantees that only signed firmware is executed during startup. Creating a trust chain starts with a read-only bootloader that validates the signature of the firmware image prior to loading using EdDSA or ECDSA.

Though secure boot provides security, it comes with its own overhead, such as flash storage requirements increasing by ~3–5 KB and a need for secure public keys storage; typically, engineers can use one-time programmable (OTP) memory.

7.5.6 KEY MANAGEMENT IN RESOURCE-CONSTRAINED ENVIRONMENTS

TinyML uses keys for authentication, encryption, and update verification. Thus, key management is one of the challenges in securing TinyML systems. Common approaches to securing key provisioning and security are:

- **Trusted Execution Environment:** Several MCUs (e.g., ARM Cortex-M33 with TrustZone) securely store the key in an isolated memory portion. However, they are not always viable or supported.
- **Key Derivation Functions (KDFs):** KDFs use master secrets and nonces to derive per-session keys. Lightweight KDFs (e.g., HKDF-SHA256) are employed in compact implementations in transport layer security (TLS), such as mbedTLS or TinyDTLS [42].
- **Pre-shared Keys (PSKs):** Such keys are hardcoded into the MCUs in manufacturing. They are simple, yet vulnerable, and lack scalability.
- **Physical Unclonable Functions (PUFs):** A new emerging hardware-based method that creates unique and device-specific keys without the need to store them. While PUFs are ideal for TinyML devices, they are still under standardization [43].

To guarantee security, key rotation and revocation should be implemented wherever feasible. However, these practices are seldom done, particularly in resource-constrained devices. In conclusion, cryptographic solutions designed to address security in TinyML need to take into consideration not only security but also efficiency and resource limitations. This section discussed how recent standards like ISO/IEC 29192 and NIST's ASCON outline feasible approaches for authentication, encryption, and firmware updates in edge device environments. Essential security methods, including MACs, key management approaches, and secure boot protocols, form a strong defense framework for TinyML systems. A promising approach ahead, including TEE, quantum-resistant algorithms, and hardware-based security techniques, would further improve protection while working under TinyML severe limitations.

7.6 PRIVACY-PRESERVING AND SECURE ML TECHNIQUES FOR TinyML

TinyML's main purpose is to transfer the power of ML from cloud-centric to devices at the edge of the network. However, transferring AI models to edge devices introduces its own novel privacy and risk threats. What if a model is compromised, stolen, or misused through adversarial inference-sensitive information can be exposed even though it hasn't been sent to the cloud. This chapter section delves into the practical methods and frameworks that improve the integrity, privacy, and confidentiality of TinyML models. It explores inference privacy protection, privacy-preserving techniques, and secure deployment procedures.

7.6.1 SECURE MODEL DEPLOYMENT

Model deployment to edge devices creates several risks including: model inversion or cloning in which the attacker can replicate the model or obtain sensitive data used during training by interacting with it and learning information from its output logit. Intellectual property (IP) leakage in which adversaries extract model binaries or reverse engineer it. Finally, model tampering, in which attackers tamper with the model weights prior to or posterior to deployment. It is essential to provide protection against such attacks or mitigate them at least. Engineers use encrypted model storage along with secure boot methods to mitigate such risks. It is essential to deploy the model in an encrypted format in flash memory and keep the key in a protected storage (e.g., OTP memory) or derive the encryption key via a secure element. Such techniques include:

- AES-128 with hardware acceleration (e.g., used in STM32Trust-like platforms)
- **Model Obfuscation:** Disguising the model weights, layers, or structures to protect against reverse engineering.

Frameworks like TensorFlow Lite for Microcontrollers (TFLM) enable the conversion of models into C arrays which are then encrypted and signed during the compilation phase in the development lifecycle. When combining that with any secure bootloader (e.g., STM32 Secure Boot or Arms's TFM), this guarantees that only trusted and genuine models are loaded and executed.

7.6.2 PRESERVING INFERENCE PRIVACY

Though TinyML is run on-device, the data being processed by the MCUs can reveal sensitive information such as user patterns and activities. For example, if a BLE message is intercepted by an adversary, they can infer sensitive data such as a person's presence in a room or a command recognized by a voice assistant to perform a specific task. The nature of the extracted information relies on the purpose of the TinyML model on the device. In order to protect the messages, methods include:

- **Secure I/O Isolation:** To reduce the leakage risk, one can encrypt the output or not provide the confidence score [44].
- Obfuscation or intermediate outputs to prevent adversaries from inferring any raw data or deducing user activity based on the responses of the model [45].

Privacy preservation frameworks such as Edge Impulse provide secure inference pipelines but reduce output granularity (e.g., providing only predicted labels and not output probabilities).

7.6.3 HOMOMORPHIC ENCRYPTION (HE) AND FEASIBILITY IN TINYML

Theoretically, homomorphic encryption (HE) enables TinyML to encrypt the input data and provide encrypted output by enabling computations on the encrypted data—raw data will not be revealed. However, such benefits come with their inherent trade-offs:

- **Long Computation Times:** processing HE-encrypted data to infer anything from them can take seconds and minutes on normal CPUs; thus, they are impractical on MCUs [46].

- **High Memory Consumption:** HE schemes (e.g., BFV or CKKS) demand megabytes of RAM, making them an infeasible option for TinyML hardware [47].

While existing HE schemes such as Helib, HEAAN, or Microsoft SEAL are computationally extensive to be implemented on real-time TinyML systems, ongoing research is studying encryption approximation schemes and hardware acceleration to enable HE to run on 32-bit embedded systems. Though HE is not yet viable for TinyML, partial HE (e.g., hybrid approaches like encrypted data offloading and edge-assisted computation) may advance with the memory and compute-improving capabilities.

7.6.4 FEDERATED LEARNING ON RESOURCE-CONSTRAINED DEVICES

In federated learning (FL), multiple participants' devices collaboratively train a global model that is shared by a central server to be trained on their local devices and data. Multiple rounds are performed, in which edge devices train the model and then share the updates with the central aggregator to form the model again. In the context of TinyML, this can preserve privacy as the data never leaves the devices. However, conventional FL implementations (e.g., TensorFlow Federated or PyTorch) are computationally heavy for edge devices. Other adaptations now exist that enable FL on MCUs:

- **FedML Lite:** An open-source framework that enables gradient sparsification and memory awareness during training for MCUs.
- **TensorFlow Lite FL:** Implements memory-mapped tensors and quantized training for edge devices.
- **Edge TPU-Based/Google Coral FL:** For federated tasks on devices with higher computation budgets (e.g., 1–2 GB RAM), not suitable for TinyML MCUs.

Still, the main challenges are communication costs and security risks, as FL demands transferring model updates at each round, which may be impractical on BLE or LPWAN. Additionally, updates are prone to poisoning by malicious clients. Despite these risks, FL is a promising, scalable approach for TinyML privacy preservation.

7.6.5 DIFFERENTIAL PRIVACY AT THE EDGE

Differential privacy (DP) ensures mathematically that the presence of any individual data point or its exclusion doesn't affect the model's output. For edge devices, local differential privacy (LDP) can be implemented by two means: first, adding small noise to the model gradients; second, performing perturbation on the model output to protect the user data [48]. This approach provides the engineers the ability to balance between the accuracy and privacy needed using ε (epsilon). Additionally, there are several lightweight schemes suitable for MCUs, such as integer Laplace noise, randomized response, and sparse vector techniques. Implementations are provided in libraries (e.g., PySyft and TFF Lite) for edge devices, though the native MCU is still evolving to support such techniques.

7.6.6 CASE STUDIES AND IMPLEMENTATIONS

Several implementations and real-world platforms are improving in privacy preservation of TinyML. Such platforms include TensorFlow Lite for Microcontrollers (TFLM), Edge Impulse, STM32 TrustZone, and PrivateFL PySyft. Table 7.3 highlights the four leading platforms offering production-ready solutions, encrypted model deployment, etc. These frameworks illustrate the trade-offs between hardware compatibility, computation overhead, and privacy assurances.

In summary, privacy-preserving approaches in TinyML systems are crucial to secure user data and model integrity, particularly with data-sensitive deployment environments such as healthcare, homes, or industry. While advanced cryptographic techniques such as HE are still impractical for most edge devices, practical techniques such as model deployment encryption, FL, and DP are already adopted in real-world TinyML systems. Figure 7.7 shows the various privacy-preserving techniques in TinyML.

Choosing the appropriate technique requires a balance between performance, communication budget, and memory budget. Previous studies summarize and compare the various privacy-preserving techniques and their resource overhead, along with framework implementation.

7.7 SECURE TINYML FRAMEWORKS, TOOLCHAINS, AND CASE STUDIES

As TinyML usage expands continuously in various real-world applications, ensuring that these application lifecycle stages are secure through the frameworks and toolchains used during that lifecycle. Each TinyML lifecycle stage, from model training and compilation to firmware flashing and inference at runtime, has its own challenging risk. This section overviews the commonly used TinyML development platforms and frameworks, exploring their provided features, and demonstrates the lessons learned from real-world applications in various applications ranging from wearable devices and smart agriculture to healthcare domains and industrial IoT.

7.7.1 SECURE TINYML FRAMEWORKS AND TOOLCHAINS

Edge Impulse is an end-to-end TinyML platform widely used in several applications; it provides signal processing, model training, deployment, and testing for MCUs. It implements several security features, including. First, secure ingestion in which user data is kept isolated per project and encrypted during transit (TLS). Secondly, digitally signed models in which model integrity and verification are ensured during compilation by a cryptographic signature during deployment. Third, built-in privacy controls in which the user's raw sensor data can be excluded from collection or cloud synchronization can be disabled to protect privacy. Edge Impulse provides secure model deployment over-the-air (OTA) using Wi-Fi or Bluetooth by cryptographic signature on firmware packages.

TensorFlow Lite for Microcontrollers (TFLM) is a lightweight framework designed for low-powered edge devices (e.g., Arm Cortex-M). While TFLM is

TABLE 7.3
Privacy-Preserving Case Studies and Frameworks

Platform	Privacy/Security Features	Hardware Compatibility	Development Status	Key Use Cases
TensorFlow Lite for Microcontrollers (TFLM)	• C-model encrypted arrays • Memory isolation (Cortex-M33/M35P) • Flash storage encryption for data protection	ARM Cortex-M, ESP32, RISC-V 59	Production-ready (Google-maintained)	TinyML devices are demanding encrypted model deployment
Edge Impulse	• Suppress raw logits/probabilities • Sensor data anonymization • RAM/Flash optimization using EON Compiler	Cross-platform (MCUs, Linux, Coral Edge TPU)	Commercial (Qualcomm-backed)	Fast prototyping of privacy-sensitive TinyML applications
STM32 TrustZone/ TrustCube-AI	• Memory protection in Arm TrustZone-based (Cortex-M33/M35P) • Firmware update validation by signature cryptographically • Encrypted flash storage (AES-256)	STM32 MCUs (e.g., WBA, MPU series)	Production-ready (STMicroelectronics)	Industrial/medical TinyML devices requiring secure boot
PrivateFL/PySyft Simulators	• Simulation of DP on FL with differential privacy • Exploratory HE researches	RP2040, Cortex-M0+ (simulated)	Research-grade (OpenMined)	Research prototyping of federated TinyML

Technique	Resource Overhead	Device Feasibility	Frameworks	Main Advantage
Encrypted Model Deployment	Low	High	TFLM, STM32	Secure model weights and IP
Inference Obfuscation	Medium	Medium	Edge Impulse	Protects against reverse inference
Homomorphic Encryption	High	Low	HEAAN, SEAL (limited)	Operates encrypted inputs directly
Federated Learning	Medium	Medium	FedML Lite, TFLM Federated	Local training, prevents raw data sharing
Differential Privacy	Low	High	TFF Lite, PySyft (simulated)	Guarantee privacy on outputs and gradients

FIGURE 7.7   TinyML privacy preservation techniques comparison.

a software framework-level, it provides security features for model building and deployment phases:

- **Model Encryption:** TFLM enables the encryption and decryption of compiled models (C array blobs) using symmetric keys at runtime.
- **Hardware Abstraction:** TFLM enables the integration of crypto co-processors and TrustZone via external libraries and build hooks.
- **Lower Attack Surface:** The small code size lowers the attack surface area and eases the integration with secure real-time OS (RTOSs).

Though TFLM doesn't provide built-in functionalities for signature verification or OTA, it is widely combined with secure bootloaders such as TF-M, STM32Cube, or MCUboot.

STM32Cube.AI is a commercial-grade toolchain provided for TinyML deployment on STM32 edge devices. It manages security via STM32Trust, combining secure boots, using TrustZone-M for memory partitioning, flash, and SRAM region encryption, and signature validation via PKI for updates.

It provides security for the model's deployment into secure memory regions, model authenticity verification at startup, and protection against runtime attacks or physical tampering.

7.7.2   SECURITY FEATURES IN COMPILERS AND DEPLOYMENT PLATFORMS

Increasingly, TinyML platforms are integrating more security features for building and runtime phases. Such features are critical for protection against model tampering, theft, or remote attacks. In real-world environments where protection measures are not implemented, systems have shown vulnerabilities to various attacks, such as model inversion, unauthorized firmware changes, etc. Previous study compares the three common platforms and their coverage for security measures. Various frameworks with their features are summarized in Figure 7.8.

Feature	Edge Impulse	TFLM	STM32Cube.AI
OTA with Signature Check	Yes	No	Yes
Inference Obfuscation	Optional	Manual	Native
Secure Boot Support	Supported	Via TF-M	Full Stack
Firmware Obfuscation	Not Native	Not Native	Optional
Tools for Data Minimization	Built-in	No	No

FIGURE 7.8   TinyML security platforms with feature coverage.

7.7.3 CASE STUDIES

a. **Industrial IoT (Predictive Maintenance):** In industry, machine vibration plays a critical role in the safety process and loosening bolted parts. It is crucial to have these machines under regular inspection for any necessary maintenance. Siemens, a German industrial company, has implemented a vibration detection TinyML on rotating machines. The system utilized STM32Cube.AI and Edge Impulse with cryptographically signed firmware distributed by Wi-Fi OTA. The system would immediately alert responsible engineers in case of a signed model validation failure, protecting the system from loading any tampered firmware [49,50].

b. **Smart Agriculture:** With scarcity of water and climate change, smart agriculture can help automate irrigation for better crops and lower water consumption. To address such agricultural challenges, a TinyML system is deployed for plant monitoring and irrigation automation. Using the Wio Terminal Microcontroller, sensors (e.g., soil moisture, temperature, humidity, and light), and Edge Impulse trained models to make a real-time decision locally. By locally processing the data collected and making decisions, the system maintains privacy and security and illustrates a viable model for low-cost and sustainable intelligent agriculture [51].

c. **Healthcare Monitoring:** Researchers have developed a TinyML-enabled system for cough detection by developing a model using Edge Impulse. The system would analyze the cough sound and the general undesired signal. The system achieved 97% accuracy. By local analysis of data and no internet access, the system guarantees the user's privacy and ensures patient anonymity [52].

7.7.4 LESSONS LEARNED FROM REAL-WORLD DEPLOYMENTS

From the so far implementation of TinyML and their privacy-preserving, security, and optimization techniques in the real world. Beneficial takeaways emerged for further improvements and design enhancements:

- The necessity of signed model updates and secure bootloaders.
- Intellectual property theft and reverse engineering can be prevented using model encryption and obfuscation.
- Digital signature checks must be enforced during OTA firmware distribution, particularly for large-scale deployments.
- In data-critical and regulated domains such as healthcare and wearables, privacy data logging and minimization are crucial for privacy and not optional.
- For real-time defenses, the integration of TinyML security frameworks and hardware security features is crucial.

All in all, securing TinyML systems is more than just model encryption—it requires a continuous validation mechanism, a secure toolchain, and a trustworthy deployment operation. In a secure deployment landscape, a complementary combination of existing tools (e.g., Edge Impulse, TensorFlow Lite Micro, and STM32Cube.AI) ensures a secure TinyML development. Real-world case studies highlight the significance of security and privacy feature integration from development to deployment in developing robust, secure TinyML systems precisely for sensitive applications such as the healthcare domain. As TinyML deployments expand to cover more sensitive applications and large-scale ones, security-by-design framework adaptation becomes vital for the security, trust, regulatory compliance, and overall success of these systems.

7.8   DESIGN GUIDELINES AND BEST PRACTICES

TinyML security and privacy is a holistic, proactive process integrating risk mitigation into design and implementation. Conventional security models can be implemented directly for TinyML due to their constraints. This section provides an outline of the best practices across TinyML related fields—design, privacy assessment, threat modeling, coding, and evaluation.

7.8.1   SECURITY-BY-DESIGN FOR TINYML

According to NIST's SP 800-160, a system security publication, incorporating protection techniques from the initial stage of the project ensures a secure system production. Implementing protective measures throughout the TinyML system development ensures minimal risk in a principle known as security-by-design. This principle is realized through:

- **Trusted Firmware and Secure Boot:** only authenticated code can be loaded and executed. Arm Trust Firmware-M (TF-M) and STM32Trust are two examples of this practice.
- **Minimal Attack Surface:** post-deployment, engineers can disable unused peripherals, debug ports, and default wireless interfaces.
- **Model Protection:** ensuring that models are encrypted and verified so that they are immune to reverse engineering or injection attacks. Precisely, when the models are in a C array format such as TensorFlow Lite Micro (TFLM).

Additionally, memory protection units (MPUs) and physical and tampering detection should be involved in a secure design, particularly in a security-critical deployment application.

7.8.2 THREAT MODELING AND RISK ASSESSMENT IN TINYML PIPELINES

Threat modeling is the process of potential risks identification and mitigation prior to deployment. To address unique TinyML threats and challenges, STRIDE (Spoofing, Tampering, Repudiation, Information disclosure, Denial of service, Elevation of privilege) can be customized for that purpose. Threats' likelihood and impact can be evaluated using tools such as OWASP Threat Dragon or Microsoft Threat Modeling Tool in which data flow is visualized and points of exposure are assessed for any risk; it can be deployed for embedded systems. The major TinyML system components and their threat exposure, along with the mitigation presented in Figure 7.9.

7.8.3 PRIVACY IMPACT ASSESSMENTS (PIAs) FOR ON-DEVICE ML

To comply with global data privacy regulations such as GDPR and HIPAA, personal data processes must be proactively assessed and clarified. Performing a Privacy Impact Assessment (PIA) aids in risk identification and the evaluation of mitigation methods. The main focus of PIA in TinyML systems should be user control, data minimization, storage and retention, and inference transparency.

Privacy threat modeling frameworks, such as LINDDUN (Linkability, Identifiability, Non-repudiation, Detectability, Disclosure of information, Unawareness, Non-compliance) can be used for risk assessment of ML systems, even in TinyML embedded systems. Figure 7.10 illustrates a privacy threat modeling divided into two main spaces: problem and solution [53].

Component	Threat	Mitigation
Sensor Interface	Replay or spoofed data	Verify timestamp, sensor fusion
ML Inference	Model inversion	Apply Output obfuscation, Suppress confidence score
OTA Firmware Update	Tampering, backdoors attack	Signed updates, version rollback protection

FIGURE 7.9 TinyML pipelines threat modeling and mitigation.

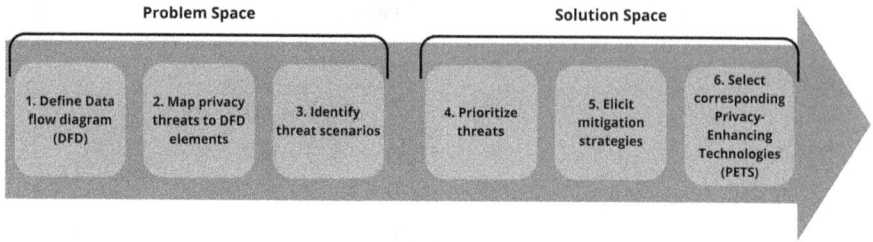

FIGURE 7.10   LINDDUN-based privacy modeling framework divided into problem and solution space [53].

7.8.4   SECURE CODING AND DEPLOYMENT PRINCIPLES

Given that C/C++ are the predominantly used programming languages in TinyML applications, they are prone to login flaws and memory errors. Secure coding standards should be adapted precisely when interacting with ML inference, such frameworks are CERT C and MISRA C. Programmers should try to avoid dynamic memory allocation (free, malloc) on MCUs, and also check device buffer boundaries prior to any input process. Additionally, they can reduce side-channel leakage by using deterministic control flow and fixed-size arrays.

Further security measures in the deployment that can enhance protection are disabling insecure update channels such as USB mass storage flashing, encrypting models' binaries (e.g., AES-128 or ASCON) if supported by the device, and deploying firmware signing and secure boot measures (e.g., MCUboot).

7.8.5   EVALUATION METRICS FOR SECURITY AND PRIVACY

For security and privacy enhancement measurement, TinyML has several metrics to evaluate the improvement, including attack surface size, model robustness, etc. Previous study highlights some of the key metrics. Standardized security and privacy benchmarking tools are still a research frontier; however, MLPerf and EdgeBench are two performance metrics projects that provide metrics such as latency and accuracy. Figure 7.11 illustrates the key evaluation metrics of TingML.

7.9   RESEARCH CHALLENGES AND FUTURE DIRECTIONS

Artificial intelligence boundaries are expanding to reach low-power edge devices by TinyML. TinyML brings. Both learning and inference to the edge. However, developing strong security and privacy practices in edge limited-resource environments is challenging operationally and technically. This section outlines the frontier research issues and emerging opportunities in academia and industry.

7.9.1   TRADE-OFFS BETWEEN ACCURACY, SECURITY, AND RESOURCE CONSUMPTION

A three-way trade-off in TinyML systems arises between security/privacy assurances, device resource limitations, and model accuracy. Large models such as CNN

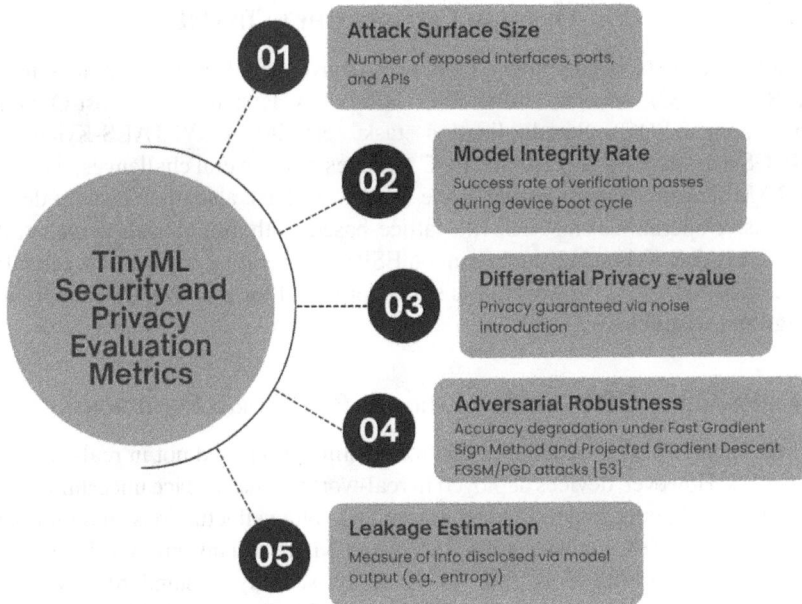

FIGURE 7.11   TinyML security and privacy evaluation metrics.

or deep neural networks (DNNs) provide higher accuracy; however, they demand larger memory and computation power, making it infeasible to run on edge devices with RAM <256 KB. Additionally, deployment of model encryption and differential privacy worsens the case as it increases the power and resource need [54,55] Model size reduction techniques, such as quantization or pruning, prove to reduce the model size at the cost of accuracy reduction by up to 10% if not carefully tuned [55]. Additionally, introduction of differential privacy reduces the model fidelity, particularly on embedded systems small datasets.

7.9.2   Advancements in Secure Model Compression and Quantization

As TinyML aims to fit on the small storage and be run in limited RAM, research is focusing on model compression methods that still retain its high performance while reducing the size and attack surface. Such techniques include integer quantization (e.g., int8), which is supported by TensorFlow Lite that eliminates floating-point units. Another technique is pruning, in which some model weights are dropped selectively while still being resistant to attacks such as backdoor or adversarial. However, several existing compression techniques are vulnerable to attacks (e.g., model inversion or extraction) in case model parameters are exposed. A privacy-first method to design models, such as encryption-aware neural architecture search (NAS), is proposed to design models with fewer leakage vectors [56].

7.9.3 FUTURE OF POST-QUANTUM CRYPTOGRAPHY IN TINYML

With advancements in quantum computing, conventional cryptographic schemes (e.g., RSA, ECC) become vulnerable. NIST has launched the Post-Quantum Cryptography (PQC) Standardization task, electing CRYSTALS-Kyber and SPHINCS+ as finalists' algorithms. PQC schemes pose several challenges, including high RAM requirements and code space (often >20 KB); also, they heavily depend on expensive operations on MCUs like lattice-based math. Initial trials demonstrated that CRYSTALS-Kyber-512 can be run on ESP332, providing hardware acceleration for AES and SHA [57]. However, it is still impractical for 8-bit MCUs used in most IoT tags and wearables.

7.9.4 PRIVACY AND SECURITY EVALUATION IN REAL-WORLD CONDITIONS

TinyML securities are evaluated in controlled simulations and not in real-world circumstances. However, devices deployed in real-world scenarios face uncertain factors such as physical tampering risk, temperature and voltage fluctuations, and unreliable connections for OTA updates. EdgeBench and MLPerf Tiny provide latency and accuracy benchmarking tools [58], yet there is a scarcity in standard privacy and security benchmarks for attacks (e.g., side-channel leakage, model inversion resistance, adversarial robustness).

7.9.5 CROSS-DISCIPLINARY OPPORTUNITIES

TinyML security is not a one teamwork. It calls for collaboration of all-related parties; MCUs engineers, ML researchers, and human-computer interaction (HCI) experts. Collaboratively, MCUs engineers work on secure design of bootloaders, memory isolation, and low-power crypto, while model resistant architecture against adversarial attacks is the work of ML researchers, finally, HCI experts can work on interface methods for edge devices with no screen or buttons for needed user privacy interactions. Emerging challenges and opportunities include the development of model auditing and explainability on MCUs [59], and Tiny federated learning for TinyML training with trust scoring [60]. These multidisciplinary opportunities bridge AI security, privacy law, and MCUs domains. While the opportunities opened by TinyML are exciting and promising, these systems security at a large scale poses significant novel challenges spanning across multiple disciplines. Future research must fill the gap between privacy, accuracy, performance and the resource constraints while being able to comply with real-world regulations. As edge intelligence spreads in everyday TinyML systems, privacy and trust by design are crucial for long-term success.

7.10 CONCLUSIONS AND REMARKS

Tiny Machine Learning (TinyML) has expanded the boundaries of AI from the central cloud to the edge of the network. This creates a new generation of smart, responsive systems in various fields such as health monitoring, industrial automation, and smart agriculture and homes. Such expansion is possible due to the deployment of optimized ML into MCUs, eliminating cloud infrastructure needs and reducing power consumption, response latency, and bandwidth usage.

As TinyML enables significant edge-based AI, its inherent distributed nature poses severe security and privacy challenges. Unlike conventional cloud ML systems, TinyML devices have no built-in isolation, large storage, or continuous network connection—such constraints make edge devices vulnerable to attacks and misuse, requiring proactive safeguards and protection. This chapter provided a foundational overview of TinyML security-related principles, such as security-by-design, threat modeling frameworks (e.g., STRIDE and LINDDUN), and architectural privacy measures like data minimization and local inference [61]. Firmware security and intellectual property protection in memory constrained devices can be protected using lightweight cryptographic mechanisms such as ASCON and model blob encryption using tools like TensorFlow Lite Micro and STM32Cube.AI. Moreover, scalable layered defense mechanisms such as federated learning and differential privacy offer promising privacy preservation and data confidentiality at edge devices [62].

Multiple real-world use cases, including on-device speech recognition by Google Coral and STMicroelectronics kits for predictive maintenance, illustrate the feasibility of TinyML security at scale. Still, TinyML security evaluation should move beyond latency and accuracy to include assessment of system robustness against adversarial attacks, privacy leakage, and model authenticity using standardized metrics [58]. Yet, challenges in deploying some techniques on 8-bit and 32-bit MCUs, such as post-quantum cryptography, still exist, and the lack of standardized security and privacy benchmarks counterparts to MLPerf Tiny hinders systems comparisons [57,63]. Along with the TinyML expansion into regulated sectors, its explainability and auditability must also mature to meet industry standards.

It is a long road ahead, and advancing TinyML requires tight collaboration among ML researchers, cryptographers, embedded system engineers, and regulatory stakeholders. Future-ready TinyML systems must incorporate privacy, security, and interpretability in development lifecycle phases from the ground up. As the domain matures, TinyML presents not just novel technical challenges but also an opportunity to establish a trustworthy AI deployment paradigm—starting from the very edge of our networks.

REFERENCES

1. P. Warden and D. Situnayake, "Tinyml: Machine learning with tensorflow lite on arduino and ultra-low-power microcontrollers," 2019. Accessed: May 19, 2025. [Online]. Available: https://books.google.com/books?hl=en&lr=&id=tn3EDwAAQBAJ&oi=fnd&pg=PP1&dq=TinyML:+Machine+Learning+with+TensorFlow+Lite+on+Arduino+and+Ultra-Low-Power+Microcontrollers&ots=jqpoeo6bB_&sig=Kzi84Xt8I0fyU0JrRkNtCMEt-P8

2. C. Banbury, C. Zhou, I. Fedorov, R. Matas, U. Thakker, D. Gope, V. Janapa Reddi, and M. Mattina, "Micronets: Neural network architectures for deploying tinyml applications on commodity microcontrollers," *Proceedings of Machine Learning and Systems*, 2021, Accessed: May 19, 2025. [Online]. Available: https://proceedings.mlsys.org/paper_files/paper/2021/hash/c4d41d9619462c534b7b61d1f772385e-Abstract.html

3. K. Reddy, B. Reddy, V. Goutham, M. Mahesh, J. S. Nisha, and G. Palanisamy. 2024, "Edge AI in sustainable farming: Deep learning-driven IoT framework to safeguard crops from wildlife threats," *ieeexplore.ieee.org*, Accessed: May 19, 2025. [Online]. Available: https://ieeexplore.ieee.org/abstract/document/10540092/

4. K. K. Hing, M. Behjati, V. Saleh, Y. K. Meng, A. P. P. A. Majeed, and Y. Zheng, "Edge intelligence for wildlife conservation: Real-time hornbill call classification using TinyML," Springer, vol. 1316 LNNS, pp. 476–488, 2025, doi: 10.1007/978–981–96–3949-6_40.

5. S. Ravi, A. Raghunathan, P. Kocher, and S. Hattangady, "Security in embedded systems: Design challenges," *ACM Transactions on Embedded Computing Systems (TECS)*, vol. 3, no. 3, pp. 461–491, Aug. 2004, doi: 10.1145/1015047.1015049.

6. J. Lin, L. Zhu, W. Chen, W.-I. Chen, and S. Han, "Tiny machine learning: Progress and futures," *IEEE Circuits and Systems Magazine*, vol. 23, pp. 8–34, 2023. Accessed: May 21, 2025. [Online]. Available: https://ieeexplore.ieee.org/abstract/document/10284551/

7. P. Warden and D. Situnayake, "Tinyml: Machine learning with tensorflow lite on arduino and ultra-low-power microcontrollers," 2019. Accessed: May 21, 2025. [Online]. Available: https://books.google.com/books?hl=en&lr=&id=t n3EDwAAQBAJ&oi=fnd&pg=PP1&dq=TinyML:+Machine+Learning+wit h+TensorFlow+Lite+on+Arduino+and+Ultra-Low-Power+Microcontrollers &ots=jqpofq6av1&sig=-atJnzJs_oPPPmEJ2fbRNn6B3x0

8. R. Immonen and T. Hämäläinen, "Tiny machine learning for resource-constrained microcontrollers," *Journal of Sensors,* vol. 2022, pp. 1–11, 2022, doi: 10.1155/2022/7437023.

9. J. Huckelberry, Y. Zhang, A. Sansone, J. Mickens, P. A. Beerel, and V. J. Reddi, "TinyML security: Exploring vulnerabilities in resource-constrained machine learning systems," Nov. 2024, Accessed: May 21, 2025. [Online]. Available: https://arxiv.org/pdf/2411.07114

10. W. Raza, A. Osman, F. Ferrini, and F. De Natale, "Energy-efficient inference on the edge exploiting TinyML capabilities for UAVs," *mdpi.com*, 2021, doi: 10.3390/drones5040127.

11. S. K. Diwaker, P. K. Chaurasia, and M. Khaliq, "Pathways for security of embedded system: Issues and challenges," *2024 2nd International Conference Computational and Characterization Techniques in Engineering and Sciences, IC3TES 2024,* 2024, Lucknow, India. doi: 10.1109/IC3TES62412.2024.10877516.

12. J. Backer, D. Hély, and R. Karri, "Secure design-for-debug for systems-on-chip," *ieeexplore.ieee.org*, 2015, Accessed: May 22, 2025. [Online]. Available: https://ieeexplore.ieee.org/abstract/document/7342418/

13. L. Batina, S. Bhasin, D. Jap, and S. Picek, "{CSI}{NN}: Reverse engineering of neural network architectures through electromagnetic side channel," *28th USENIX Security Symposium (USENIX Security 19), 2019usenix.org*, Santa Clara CA, USA. Accessed: May 22, 2025. [Online]. Available: https://www.usenix.org/conference/usenixsecurity19/presentation/batina

14. S. Maji, U. Banerjee, and A. P. Chandrakasan, "Leaky nets: Recovering embedded neural network models and inputs through simple power and timing side-channels— Attacks and defenses," *IEEE Internet of Things Journal*, vol. 8, no. 15, pp. 12079–12092, Aug. 2021, doi: 10.1109/JIOT.2021.3061314.

15. A. Rakin, Z. He, and D. Fan, "Bit-flip attack: Crushing neural network with progressive bit search," *Proceedings of the IEEE/CVF International Conference on Computer Vision (ICCV)*, 2019, Seoul, Korea (South). Accessed: May 22, 2025. [Online]. Available: https://openaccess.thecvf.com/content_ICCV_2019/html/Rakin_Bit-Flip_Attack_Crushing_Neural_Network_With_Progressive_Bit_Search_ICCV_2019_paper.html

16. T. Perkovic, J. Sabic, K. Zovko, and P. Solic, "An investigation of a replay attack on LoRaWAN wearable devices," *2023 IEEE International Mediterranean Conference on Communications and Networking, MeditCom 2023*, Dubrovnik, Croatia. pp. 45–49, 2023, doi: 10.1109/MEDITCOM58224.2023.10266648.

17. "(PDF) MitM attacks and IoT security: A case study on MQTT," Accessed: May 22, 2025. [Online]. Available: https://www.researchgate.net/publication/376558974_MitM_Attacks_and_IoT_Security_A_Case_Study_on_MQTT

18. M. Fredrikson, S. Jha, and T. Ristenpart, "Model inversion attacks that exploit confidence information and basic countermeasures," *Proceedings of the ACM Conference on Computer and Communications Security*, pp. 1322–1333, Oct. 2015, Denver Colorado, USA. doi: 10.1145/2810103.2813677

19. J. Huckelberry, Y. Zhang, A. Sansone, J. Mickens, P. A. Beerel, and V. J. Reddi, "TinyML security: Exploring vulnerabilities in resource-constrained machine learning systems," Nov. 2024, Accessed: May 23, 2025. [Online]. Available: https://arxiv.org/pdf/2411.07114

20. J. Michael, "Security and privacy for edge artificial intelligence," *computer.org*, Accessed: May 23, 2025. [Online]. Available: https://www.computer.org/csdl/magazine/sp/2021/04/09475182/1uZskexqBJ6

21. R. Sachdeva and S. Bhatia, "Security and privacy in edge AI: Challenges and concerns," In R. Sachdeva and S. Bhatia (Eds.), *Edge Computational Intelligence for AI-Enabled IoT Systems*, pp. 69–99, Jan. 2024, CRC Press. doi: https://www.taylorfrancis.com/chapters/edit/10.1201/9781032650722.

22. N. Chawla and S. Dalal, "Edge AI with wearable IoT: A review on leveraging edge intelligence in wearables for smart healthcare," In S. Dalal, V. Jaglan, and D.-N. Le (Eds.), *Green Internet of Things for Smart Cities*, pp. 205–231, Jun. 2021, CRC Press. doi: https://www.taylorfrancis.com/chapters/edit/10.1201/9781003032397-14.

23. P. Cheng and U. Roedig, "Personal voice assistant security and privacy—A survey," *ieeexplore.ieee.org*, 2022, Accessed: May 23, 2025. [Online]. Available: https://ieeexplore.ieee.org/abstract/document/9733178/

24. K. Chen, Y. Lin, H. Luo, B. Mi, Y. Xiao, C. Ma, and Jorge Sá Silva, "Edgeleakage: Membership information leakage in distributed edge intelligence systems," *arxiv.org*, Accessed: May 23, 2025. [Online]. Available: https://arxiv.org/abs/2404.16851

25. C. Dwork and A. Roth., "The algorithmic foundations of differential privacy," *Foundations and Trends® in Theoretical Computer Science, 2014•nowpublishers.com*, doi: 10.1561/0400000042.

26. P. Qi, D. Chiaro, and F. Piccialli, "Small models, big impact: A review on the power of lightweight Federated Learning," *Future Generation Computer Systems*, vol. 162, p. 107484, Jan. 2025, doi: 10.1016/J.FUTURE.2024.107484.

27. W. Zhu, M. Goudarzi, and R. Buyya, "FLight: A lightweight federated learning framework in edge and fog computing," *Software: Practice and Experience,* vol. 54, no. 5, pp. 813–841, May 2024, doi: 10.1002/SPE.3300.

28. L. Zhu, Z. Liu, and S. Han, "Deep leakage from gradients," *Advances in Neural Information Processing Systems*, vol. 32, pp. 14774–14784, 2019.

29. E. Bagdasaryan, A. Veit, Y. Hua, D. Estrin, V. Shmatikov, and C. Tech, "How to backdoor federated learning," Jun. 03, 2020, *PMLR*. Accessed: May 23, 2025. [Online]. Available: https://proceedings.mlr.press/v108/bagdasaryan20a.html

30. C. Chhetri and V. G. Motti, "User-centric privacy controls for smart homes," *Proceedings of the ACM on Human-Computer Interaction*, vol. 6, no. 2 CSCW, Nov. 2022, doi: 10.1145/3555769.

31. G. Annas, "HIPAA regulations: A new era of medical-record privacy?" *scholarship. law.bu.edu*, 2003, Accessed: May 23, 2025. [Online]. Available: https://scholarship.law. bu.edu/faculty_scholarship/1283/?trk=public_post_comment-text

32. E. GDPR, "General data protection regulation (gdpr)," 2018, Accessed: May 23, 2025. [Online]. Available: https://in2mobile.gr/wp-content/uploads/2018/05/GDPR_ NEWSurvey.pdf

33. E. Bryan, "The culture of surveillance: Watching as a way of life," *Social & Cultural Geography*, vol. 19, pp. 1107–1109, Sep. 2018, doi: 10.1080/14649365.2018.1525837.

34. P. Schwabe and K. Stoffelen, "All the AES you need on Cortex-M3 and M4," *International Conference on Selected Areas in Cryptography*, vol. 10532 LNCS, Ottawa, Canada. pp. 180–194, 2017, doi: 10.1007/978-3-319-69453-5_10.

35. A. Hassan, "Lightweight cryptography for the Internet of Things," *Proceedings of the Future Technologies Conference (FTC)*, vol. 1290, pp. 780–795, 2021, doi: 10.1007/978-3-030-63092-8_52.

36. "Lightweight cryptography," *CSRC*, Accessed: May 24, 2025. [Online]. Available: https://csrc.nist.gov/Projects/lightweight-cryptography.

37. "Information security-lightweight cryptography—Part 2: Block ciphers International Standard ISO/IEC 29192-2 Copyright Protected Document," Accessed: May 24, 2025. [Online]. Available: www.iso.org

38. R. Beaulieu, S. Treatman-Clark, D. Shors, B. Weeks, J. Smith, and L. Wingers, "The SIMON and SPECK lightweight block ciphers," *DAC '15: Proceedings of the 52nd Annual Design Automation Conference*, June 7–11, 2015, doi: 10.1145/2744769.2747946.

39. N. Mouha, B. Mennink, A. Van Herrewege, D. Watanabe, B. Preneel, and I. Verbauwhede, "Chaskey: An efficient MAC algorithm for 32-bit microcontrollers," *Selected Areas in Cryptography—SAC 2014: 21st International Conference,* Montreal, QC, Canada. vol. 8781, pp. 306–323, 2014, doi: 10.1007/978-3-319-13051-4_19.

40. D. J. Bernstein, N. Duif, T. Lange, P. Schwabe, and B.-Y. Yang, "High-speed high-security signatures," *Journal of Cryptographic Engineering*, vol. 2, no. 2, pp. 77–89, Sep. 2012, doi: 10.1007/S13389-012-0027-1.

41. B. Moran, H. Tschofenig, D. Brown, and M. Meriac, "A firmware update architecture for Internet of Things," Apr. 2021, doi: 10.17487/RFC9019.

42. T. Fossati, "RFC 7925: Transport Layer Security (TLS)/Datagram Transport Layer Security (DTLS) profiles for the Internet of Things," Jul. 2016, doi: 10.17487/RFC7925.

43. A. Al-Meer and S. Al-Kuwari, "Physical Unclonable Functions (PUF) for IoT devices," *ACM Comput Surv*, vol. 55, no. 14 S, Dec. 2023, doi: 10.1145/3591464.

44. P. Mahadevappa, R. K. Murugesan, R. Al-amri, R. Thabit, A. H. Al-Ghushami, and G. Alkawsi, "A secure edge computing model using machine learning and IDS to detect and isolate intruders," *MethodsX*, vol. 12, p. 102597, Jun. 2024, doi: 10.1016/J. MEX.2024.102597.

45. A. Munir, E. Blasch, J. Kwon, J. Kong, and A. Aved, "Artificial intelligence and data fusion at the edge," *ieeexplore.ieee.org*, Accessed: May 25, 2025. [Online]. Available: https://ieeexplore.ieee.org/abstract/document/9475883/

46. I. Chillotti, N. Gama, M. Georgieva, and M. Izabachène, "TFHE: Fast fully homomorphic encryption over the torus," *Journal of Cryptology,* vol. 33, no. 1, pp. 34–91, Jan. 2020, doi: 10.1007/S00145-019-09319-X.

47. "Microsoft SEAL (Simple Encrypted Arithmetic Library)—Google scholar," Accessed: May 25, 2025. [Online]. Available: https://scholar.google.com/scholar?hl=en&as_sdt=0% 2C5&q=Microsoft+SEAL+%28Simple+Encrypted+Arithmetic+Library%29&btnG=

48. C. Dwork and A. Roth, "The algorithmic foundations of differential privacy," *Foundations and Trends® in Theoretical Computer Science*, vol. 9, no. 3–4, pp. 211–407, 2014, doi: 10.1561/0400000042.

49. I. B. Demir, "Artificial Intelligence for predictive maintenance," 2023, Accessed: May 26, 2025. [Online]. Available: https://www.politesi.polimi.it/handle/10589/222753

50. E. Oye, O. Emma, M. Luther, and S. Emerson, "Artificial intelligence in utilities: Predictive maintenance and beyond," *researchgate.net*, Accessed: May 26, 2025. [Online]. Available: https://www.researchgate.net/profile/Satyaveda-Somepalli/publication/389783192_Artificial_Intelligence_in_Utilities_Predictive_Maintenance_and_Beyond/links/67d20267e62c604a0dd754f9/Artificial-Intelligence-in-Utilities-Predictive-Maintenance-and-Beyond.pdf

51. V. Tsoukas, A. Gkogkidis, and A. Kakarountas, "A TinyML-based system for smart agriculture," doi: 10.1145/3575879.3575994.

52. A. Rana, Y. Dhiman, and R. Anand, "Cough detection system using TinyML," *Proceedings -2022 International Conference on Computing, Communication and Power Technology, IC3P 2022*, Visakhapatnam, India. pp. 119–122, 2022, doi: 10.1109/IC3P52835.2022.00032.

53. A. Robles-González, J. Parra-Arnau, and J. Forné, "A LINDDUN-based framework for privacy threat analysis on identification and authentication processes," *Computers & Security*, vol. 94, p. 101755, Jul. 2020, doi: 10.1016/J.COSE.2020.101755.

54. N. Alajlan and D. Ibrahim, "TinyML: Enabling of inference deep learning models on ultra-low-power IoT edge devices for AI applications," *mdpi.com*, Accessed: May 26, 2025. [Online]. Available: https://www.mdpi.com/2072-666X/13/6/851

55. S. Han, H. Mao, and W. J. Dally, "Deep compression: Compressing deep neural networks with pruning, trained quantization and Huffman coding," *4th International Conference on Learning Representations, ICLR 2016—Conference Track Proceedings*, 2016, San Juan, Puerto Rico. Accessed: May 26, 2025. [Online]. Available: https://arxiv.org/pdf/1510.00149

56. Z. Ghodsi, A. Veldanda, B. Reagen, and S. Garg, "Cryptonas: Private inference on a relu budget," *proceedings.neurips.cc*, Accessed: May 26, 2025. [Online]. Available: https://proceedings.neurips.cc/paper/2020/hash/c519d47c329c79537fbb2b6f-1c551ff0-Abstract.html

57. F. Segatz, M. Ihsan, and A. Hafiz, "Efficient implementation of CRYSTALS-KYBER key encapsulation mechanism on ESP32," Mar. 2025, Accessed: May 26, 2025. [Online]. Available: https://arxiv.org/pdf/2503.10207v1

58. C. Banbury et al., "MLPerf tiny benchmark," Jun. 2021, Accessed: May 26, 2025. [Online]. Available: https://arxiv.org/pdf/2106.07597

59. B. Kim, M. Wattenberg, J. Gilmer, C. Cai, J. Wexler, and F. Viegas, "Interpretability beyond feature attribution: Quantitative testing with concept activation vectors (tcav)," *proceedings.mlr.pressInternational conference on machine learning, 2018proceedings.mlr.press*, 2018, Accessed: May 26, 2025. [Online]. Available: https://proceedings.mlr.press/v80/kim18d.html

60. "FedProx: Federated optimization in heterogeneous networks—Flower baselines 1.19.0," Accessed: May 26, 2025. [Online]. Available: https://flower.ai/docs/baselines/fedprox.html

61. K. Malhotra and Y. Kumar, "Challenges to implement machine learning in embedded systems," *Proceedings—IEEE 2020 2nd International Conference on Advances in Computing, Communication Control and Networking, ICACCCN 2020*, Greater Noida, India. pp. 477–481, Dec. 2020, doi: 10.1109/ICACCCN51052.2020.9362893.

62. K. Bonawitz, H. Eichner, W. Grieskamp, D. Huba, A. Ingerman, V. Ivanov, C. Kiddon, and J. Konečný, "Towards federated learning at scale: System design," *Proceedings of Machine Learning and Systems, 2019proceedings.mlsys.org*, 2019, Accessed: May 26, 2025. [Online]. Available: https://proceedings.mlsys.org/paper_files/paper/2019/hash/7b770da633baf74895be22a8807f1a8f-Abstract.html

63. C. R. Banbury et al., "Benchmarking TinyML systems: Challenges and direction," Mar. 2020, Accessed: May 26, 2025. [Online]. Available: https://arxiv.org/pdf/2003.04821

8 Secure Tiny Machine Learning on Resource-Constrained IoT Devices

Yassine Maleh, Ismail Lamaakal,
and Khalid EL Makkaoui

8.1 INTRODUCTION

The advent of the Internet of Things (IoT) marks a decisive turning point in the evolution of embedded systems and communication technologies. This revolutionary transformation not only heralds a transition to a more connected technology ecosystem, but it also introduces a host of new complexities that are redefining traditional security paradigms. The Internet has expanded far beyond a simple network of computers to encompass a wide range of devices, including RFID tags, sensor networks, automotive networks, and many other interconnected technologies.

At the heart of this revolution is the Internet of Things, which involves the networking of many appropriate wireless entities and devices, such as mobile phones and sensors, made possible by various communication protocols. This level of interconnection applies to objects with significant practical consequences, thus requiring a robust security architecture to reduce risk. Therefore, protocols need to be revised to incorporate new limitations, focusing on strengthening security mechanisms.

In this rapidly changing context, Tiny Machine Learning (TinyML) is emerging as a critical area at the forefront of this transformation. TinyML represents the intersection between machine learning and embedded systems, enabling the execution of artificial intelligence models on devices with microcontrollers with very low-power consumption. This technological convergence opens up new possibilities for intelligent and autonomous operations at the edge of networks, potentially revolutionising multiple industries, including healthcare, manufacturing, and environmental monitoring.

TinyML represents a revolutionary approach to edge computing that explores the deployment and training of machine learning models directly on edge devices. This technology bridges the gap between resource-constrained IoT devices and edge computing capabilities, enabling machine learning applications on ultra-low-power devices.

The transformation of the technology landscape is particularly visible in the evolution of the TinyML Foundation, which now identifies itself as the EdgeAI Foundation. This evolution reflects the expansion of the field: while TinyML initially

DOI: 10.1201/9781003544449-8

focused on running machine learning models on small devices with limited processing power and memory, EdgeAI extends this vision by incorporating more complex models and using local computing resources for real-time decision-making.

The applications of the edge analytics market are transforming industries, and the edge computing market is expected to reach around $350 billion by 2027. However, the current approach to edge analytics involves machine learning models trained in the cloud, which introduces latency into the system and is prone to privacy concerns. TinyML offers a new approach to edge computing that explores the deployment and training of machine learning models directly on edge devices.

Despite the promising advancements of TinyML, this technology faces unique and unprecedented security challenges. TinyML devices operate with strict resource limitations, possessing memory and computational power that are often two to three orders of magnitude lower than traditional IoT or edge devices. This significant disparity in resources introduces unprecedented challenges for implementing security measures.

While a standard edge device might have megabytes of RAM and processors running at gigahertz speeds, TinyML devices typically operate with only kilobytes of memory and processors operating at megahertz speeds. These severe constraints make it impractical, if not impossible, to apply conventional safety methods directly. In addition, the wide deployment of these resource-constrained devices in diverse and often physically accessible environments exposes them to a variety of potential threats.

The distinguishing features of TinyML devices present a unique set of security concerns that require careful consideration. Traditional security approaches, designed for resource-intensive environments, often conflict with the strict limitations of TinyML systems. This mismatch leads to significant vulnerability in protecting these devices from various threats.

The importance of this research is underscored by the significant gap between the rapid advancements of TinyML technologies and the relatively slow pace of research on their security. An analysis of release trends reveals an alarming disparity: between 2015 and 2023, about 347 publications were dedicated to TinyML models, hardware, and software, but only nine addressed the issue of TinyML security.

This statistic reveals a critical gap in academic and industrial research. Even more concerningly, many of the security-focused articles focus more on using TinyML to improve security rather than addressing the security of the TinyML devices themselves. This lack of security research, despite the increasing implementation of TinyML in various fields, underscores the urgent need for in-depth review and analysis.

This paper aims to fill this critical gap by providing a comprehensive and organised assessment of the security landscape within TinyML. The main objectives of this chapter is to develop a comprehensive taxonomy of edges, distinguishing between traditional IoT devices, EdgeML devices, and TinyML devices, elucidating the distinct security challenges each type faces. This paper mainly aims to make the following contributions:

- It defines various requirements of a good threat and mitigation classification.
- It analyses the constrained IoT security threats.
- It studies and analyses in-depth key management protocols and TinyML.

The structure of this chapter is as follows. Section 8.2 describes the research methodology. Section 8.3 reviews the related constrained IoT requirements and threats. Section 8.4 discusses security threats. Section 8.5 presents the most important key management schemes for constrained IoT. Finally, Section 8.6 concludes the chapter.

8.2   RESEARCH METHODOLOGY

We delve into pivotal aspects of security in resource-constrained IoT environments, with a particular focus on Tiny Machine Learning (TinyML) systems. This inquiry addresses fundamental questions related to security requirements, threat models, and key management schemes, aiming to illuminate critical elements necessary to ensure the integrity, confidentiality, and resilience of intelligent edge devices.

Understanding the evolving security landscape of TinyML-enabled IoT systems is essential, as these devices operate under strict limitations in terms of processing power, memory, and energy, making traditional security approaches ineffective. We analyse the primary security requirements, including data integrity, confidentiality, authentication, and lightweight privacy mechanisms, which are essential to maintain trust, especially in sensitive and dynamic edge computing environments.

We also examine common threats such as model inversion attacks, adversarial inputs, data exfiltration, physical tampering, and unauthorised access, all of which pose significant risks to the deployment of TinyML in real-world scenarios. These threats highlight the pressing need for adaptive and lightweight security mechanisms.

Furthermore, we explore various key management schemes tailored to constrained devices, emphasising efficient key generation, distribution, storage, and renewal mechanisms compatible with TinyML's computational limitations. By evaluating existing approaches, we identify best practices, current limitations, and open challenges in securing learning-enabled IoT ecosystems.

Through this comprehensive analysis of security requirements, emerging threats, and key management strategies, this section contributes critical insights into the design of secure TinyML frameworks, paving the way for robust, scalable, and context-aware security solutions for the next generation of intelligent embedded systems (Maleh et al. 2022).

Here we detail the procedures that were followed to conduct this survey. We depend on the process of Systematic Literature Reviews (SLRs) to enhance our understanding of related categories. SRL supposedly differentiates itself from conventional common reviews by pursuing a reproducible, scientific, and transparent procedure. Initially implemented in the medical profession, SLRs offer a reproducible research approach and should yield enough data for other researchers to duplicate. Several researchers in the field of computer security have recently begun employing the SRL technique. With the emergence of high-quality research on SDN security, the SLR can help classify these studies, which are still in their early phases.

The research topics were defined initially to organise the SRL. The potential benefits of TINYML-ENABLED DEVICES security had to be considered for this. This established three foundational subject categories: exposures, assaults, and critical management processes. To emphasise the need to conduct a current and comprehensive assessment of TinyML-enabled devices' risks and important management strategies, we have posed two research questions that we shall endeavour to answer in this paper:

RQ1: What are the main security requirements and threats in resource-constrained IoT environments enabled by Tiny Machine Learning?

RQ2: What are the key management schemes proposed for securing TinyML-enabled systems and low-power IoT devices?

We use a three-stage selection approach to select the articles pertaining to these research issues: (1) defining search keywords, (2) selecting sources, and (3) determining and applying inclusion/exclusion criteria for chosen papers:

1. **Search Terms:** This step focuses on identifying the most relevant keywords for evaluating security challenges and mitigation strategies in the context of Tiny Machine Learning (TinyML) on resource-constrained IoT devices. To ensure a comprehensive literature search, we included core terms along with relevant synonyms and variations. Selected keywords include:
   "TinyML security," "Security in Tiny Machine Learning," "IoT device vulnerabilities," "Lightweight security protocols," "Secure TinyML," "Key management in IoT," "Security for resource-constrained devices," "Edge AI security," "Threats in TinyML systems," and "TinyML-enabled IoT attacks."

2. **Search String Definition:** To define the search string, we used Boolean operators—"OR" to group synonyms and related terms, and "AND" to combine key concepts such as TinyML, security, and constrained IoT systems. The resulting search string is
   (("Tiny Machine Learning" OR "TinyML" OR "Edge AI" OR "Embedded ML") AND ("IoT" OR "Internet of Things" OR "constrained devices" OR "resource-limited systems") AND ("security" OR "secure" OR "attacks" OR "vulnerabilities" OR "threats" OR "lightweight protocols" OR "key management"))

3. **Selection of Sources:** For this search phase, we have chosen to use the popular digital academic databases listed in Table 8.1. For this review, we looked at both journal publications and conference proceedings. Some of

TABLE 8.1

Digital Databases Used in the SRL

Online Databases #	Online Database	URL
1	IEEE	http://ieeexplore.ieee.org/
2	ScienceDirect	http://www.sciencedirect.com/
3	ACM	http://www.acm.org/
4	Wiley	http://www.wiley.com/
5	SpringerLink	http://link.springer.com/
6	Hindawi	https://www.hindawi.com/
7	Taylor and Francis	http://taylorandfrancis.com/

TABLE 8.2

Summary of the Inclusion-Exclusion Criteria for Selection Papers

Criterion	Rationale
Inclusion 1	Early-stage and recent studies proposing key management schemes specifically designed for resource-constrained IoT devices are included, as they address the unique security challenges posed by limited computational and energy capabilities.
Inclusion 2	Studies that define or analyse security requirements, threats, and attack models in TinyML or embedded IoT systems are considered to ensure a broad understanding of the constrained device security landscape.
Inclusion 3	Research exploring security architectures or lightweight cryptographic protocols in the context of TinyML or secure edge intelligence is included to support the development of secure learning at the edge.
Exclusion 1	Studies that do not focus on security aspects in TinyML, constrained IoT, or edge ML environments are excluded, as the scope is limited to securing ML on low-power devices.
Exclusion 2	Non-English publications are excluded to maintain consistency and accessibility in the literature review process.
Exclusion 3	Studies published before 2010 or after 2024 are excluded, with additional screening during full-text analysis to ensure relevance to TinyML-related security and key management topics.

the papers that were downloaded had their references checked, and we also added publications that were relevant to the subject. For various reasons, such as the difficulty or infeasibility of instantiating the search string, the high index of duplicate articles with IEEE and ScienceDirect databases, or the fact that the databases (like Google Scholar) only performed one indexing, other databases were not utilised.

4. **Inclusion/Exclusion Criteria:** Table 8.2 summarises the inclusion-exclusion criteria for the SRL (Figure 8.1).

8.3   SECURITY REQUIREMENT IN TinyML

To protect the authenticity, privacy, and secrecy of patient information, it is essential to fulfil the following fundamental security needs:

Data Originality: Must be met by TinyML-enabled devices to guarantee the security and dependability of transmitted data. To do this, services related to licencing and verification are required (Yassine and Ezzati 2015). Each sensor and base station in a TINYML-ENABLED DEVICE must use its authentication technique to ensure genuine data. This ensures the data is trustworthy and accurate, which helps avoid wrong diagnoses and treatments. Another important goal of authentication techniques is ensuring that patient data remains discreet and private throughout transmission (Maleh et al. 2016).

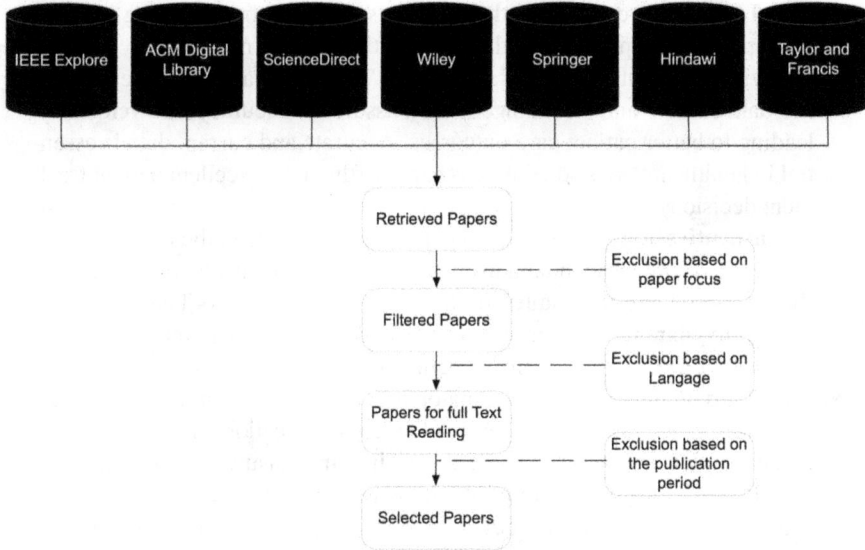

FIGURE 8.1   The proposed papers' selection methodology.

Data Privacy: Attackers can eavesdrop on sensitive data being shared between nodes in TinyM-enabled devices because of the wireless channel. The end consequence can be the unauthorised disclosure of private patient information. One solution to this problem is to encrypt data before transferring it. Only approved users will be able to decipher this encrypted material. Employing encryption methods is crucial in ensuring that sensitive patient information remains secure from unauthorised parties. The data will be protected during transmission because of this.

Data Integrity: Important for protecting sensitive patient information during transmission. System failures and patient harm can result from intruders intercepting and altering data. This highlights the need for safeguards to avoid data manipulation while in transit. Data integrity checks, which ensure the data received at the destination is real, are one approach to doing this. These verifications examine if the data that has been received has been altered or changed in any way by comparing it to the data that was originally transmitted. Healthcare professionals may make better decisions regarding patient care when they have access to accurate and trustworthy patient data, which is why data integrity measures are so important.

Data Availability: Medical sensor nodes must be readily available to guarantee that health data is always available for medical treatment. Data loss may occur if an unauthorised person were to seize a sensor node, making it very difficult to provide medical treatment. As a result, keeping medical care software accessible is crucial for ensuring that doctors and other medical staff have constant access to vital patient records.

Updated Data: A freshness method prevents attackers from reusing old data by implementing protections that prevent the attacker node from recording, replaying, and publishing data. Because TinyML-enabled devices ensure the data is new and undamaged, they assure its accuracy and reliability, leading to better patient care outcomes. Accurate and current data is essential in healthcare apps so clinicians can swiftly make excellent patient treatment decisions.

Data Authentication: Every node may identify and confirm the identity of the nodes that supply it with data using an authentication mechanism, ensuring the data is genuine and unaltered. Data authentication allows TinyM-enabled devices to guarantee that only authorised nodes may access and share sensitive patient data, which improves patient privacy and security.

Secure Management: The coordinator has a number of options for safely distributing keys, including encrypted communication channels, secure authentication mechanisms, and other similar measures. In addition, the coordinator can take the required steps to revoke the keys to stop unauthorised individuals from accessing the network and its data. Using a coordinator and secure key management procedures allows WBANs to better safeguard sensitive data during transmission and reception on the network, preventing unauthorised access.

Dependability: For the receiver to be able to identify and fix any transmission problems, error coding is a method that includes adding redundant data to the sent data. If the data being communicated is susceptible to noise or interference from the wireless channel, this method can be very effective in TinyML-enabled devices. Regarding applications like medical monitoring, where data mistakes might have life-altering effects, TinyML-enabled devices' error coding feature is invaluable for ensuring the data supplied and received by the network is accurate and reliable. Further, by decreasing the possibility of data corruption or loss caused by transmission mistakes, error coding can assist in making the network more reliable.

Safe Positioning: intruders can compromise patient privacy and security by entering fraudulent signals and information into the location registration system through patient movements and updates. Avoiding this problem is possible with the use of secure authentication and encryption techniques, which limit the ability to update the patient's location information to authorised devices only. In addition, the location registration system may be set up with frequent monitoring and detection systems to catch suspicious activities or attempts at unauthorised access. To protect the confidentiality of patient information and stop unauthorised people from accessing or tampering with their location data, patients must be securely positioned in WBANs. With the right safeguards in place, WBANs can ensure that patient location data is tracked safely and reliably.

Accountability: All healthcare team members, including administrators, should be aware of the gravity of the situation and act accordingly to keep patients' personal information secure. Someone must be held accountable if they are responsible for the unauthorised use or disclosure of patient information, because it can have catastrophic implications. Consequently, it is critical to set up transparent protocols for handling and securing patient information and to check in on these steps often to ensure they're working.

Flexibility: Patients may need a second party or hospital to access their information in the event of an emergency. To guarantee the privacy and accuracy of the data, the system should offer a safe and effective way for authorised parties to access and share this information. The collection of patient permission and the regulation of data access are two of the most important components of an effective policy and procedural framework for sharing patient information.

Privacy and Compliance: Ensuring the security of patients' private information is the highest priority. Worldwide standards and legislation, such as the US's Health Insurance Portability and Accountability Act (HIPAA), have been implemented to guarantee this. Penalties, both criminal and civil, such as fines and jail time, await those who disobey these rules. Healthcare providers must follow these rules and protect their patients' personal information.

Data Authenticity: Data must be legitimate and originated from reliable sources for wireless medical sensor networks to function properly. One can employ authentication methods such as public and private keys to do this. The data is encrypted using public keys and decrypted using private keys. This ensures that no unauthorised parties may access the data and that it hasn't been altered. The trustworthiness and precision of patient records, which are essential for their care and treatment, depend on data authentication.

Data Authorisation: The authorisation technique controls user access to network resources and services. Access policies and access control lists (ACLs) work together to provide fine-grained management of who has access to what on a network. In healthcare settings, where unauthorised access to sensitive patient data is a major concern, this is of the utmost importance.

8.4   SECURITY THREATS IN TiNyM-ENABLED DEVICES

The monitoring and tracking of vital signs in body sensor networks raises serious concerns about patient privacy. The unauthorised exposure of sensitive patient information might occur if attackers eavesdrop on communication lines. An adversary with access to a powerful enough receiver antenna can intercept a patient's network connections and steal sensitive information such as their position, time stamps, message IDs, and source and destination addresses. Malicious actors can utilise this data to inflict bodily harm. Patients' right to privacy and security are gravely endangered by these kinds of eavesdropping operations (Butt et al. 2019). Medical Internet of Things (IoT) sensor systems often employ wireless networks for communication, which have vulnerabilities during transmission and are not intrinsically restricted in communication range. This might lead to possible dangers to sensitive data. The information that medical IoT sensors communicate to the doctor's and hospital's servers may be tampered with if an attacker were to intercept and alter the data. An extremely dangerous situation may arise if this unauthorised data tampering includes changing the patient's physiological data and sending it to a server. It is crucial to prioritise implementing strong security measures to safeguard patients' data during transmission, ensuring its confidentiality, integrity, validity, and privacy (Abouzakhar et al. 2017).

8.4.1   DIFFERENT TYPES OF ATTACKS

Interception: The wireless communication route that carries the patient's vital signs data from their device to their healthcare practitioner becomes compromised when an attacker obtains access. Once intercepted, this data might contain sensitive information, including a patient's location, medical history, or other personal details.

Message Change: This attack occurs when someone or something other than the intended receiver can intercept a communication and change its contents before sending it. In healthcare settings, this kind of assault is especially problematic since it might cause the transmission of false medical instructions or information, which can have a negative impact on the patient's health. An attacker may, for example, tamper with a patient's prescription, causing them to experience unwanted side effects. Hence, methods like digital signatures and encryption are crucial for healthcare communication to guarantee the authenticity and integrity of messages.

Wireless Sensor Routing Threats: In the context of wireless sensor networks, these terms describe harmful actions taken at the network layer. False alerts might be caused by these threats' activities, which include stealing or altering packets and sending them to the remote control centre. Another way attackers might cause network disruption is by tampering with the address fields of captured packets. This can lead to routing loops or even the complete network disruption. Transmitted data may lose its integrity and confidentiality due to these assaults. As a result, protecting wireless sensor networks against routing attacks requires strong security measures.

8.4.1.1   Active Attacks on TinyML-Enabled Devices

Because they can jeopardise the data's validity, integrity, and secrecy, active assaults pose a major threat to WBSNs and other wireless sensor networks. To cause confusion and damage, the attacker can alter the data, insert bogus data, or play back previously recorded data. Moreover, the attacker can take over the network without anyone's knowledge by masquerading as a valid node and stealing important data.

DoS Attack: The goal of a denial-of-service (DoS) assault is to block access to a system or network by many legitimate users by flooding it with traffic or other harmful activities. Specifically, DoS attacks can interrupt the regular operation of wireless body area networks (WBANs), which hinders the transmission of medical data and might put the patient at risk. The "jamming" attack is a common DoS technique against WBANs. This kind of attack involves the perpetrator overwhelming the wireless spectrum with excessive noise or interference, impeding or blocking the transmission of valid data. Serious health complications may arise if the medical personnel overseeing a patient did not get vital signs or other data collected by the patient's sensors. A further denial-of-service attack in WBANs is known as a "sleep deprivation" assault. This kind of assault involves the perpetrator repeatedly sending wake-up signals to the patient's battery-powered sensors, stopping them from going into power-saving low-power sleep modes. This can rapidly deplete the battery life of the sensors, making them unable to transmit critical medical data. Medical monitoring systems are generally quite vulnerable to denial-of-service assaults in WBANs (Sikder et al. 2021).

8.4.1.1.1 *Physical Attacks*

Because they are dispersed and vulnerable, physical attacks pose a significant risk to outdoor wireless networks. Compared to wired networks, they are more prone to these types of assaults. Physical attacks on sensor nodes are common and often result in permanent damage. Attackers can steal sensitive information or alter software code to do more damage if they gain physical access.

8.4.1.1.2 *Deceptive Routing Information*

This attack hopes to cause performance degradation by rerouting network traffic by forming routing loops or less-than-ideal pathways. An adversary can alter data packet header information during transmission to provide erroneous routing information. One can play with routing metrics such as hop counts, source or destination addresses, and so on to achieve this. This attack can spread throughout the network if additional nodes get this erroneous routing information and utilise it to route their packets.

8.4.1.1.3 *Blackhole Attack*

Here, a bad actor in the network makes an exaggerated claim about the quickest route to the sink node—the final destination of the network's gathered data. The malicious actor will start ignoring or discarding any data packets sent to it after convincing other network nodes that it has the shortest path (Mucchi et al. 2019). Consequently, the network is severely compromised since all packets destined for the sink node are lost. There are a number of strategies at the blackhole attacker's disposal for making other nodes believe it has the shortest path. For instance, it can impersonate routing messages, alter the hop count or sequence number of routing messages, or take advantage of network protocol vulnerabilities, as shown in Figure 8.2.

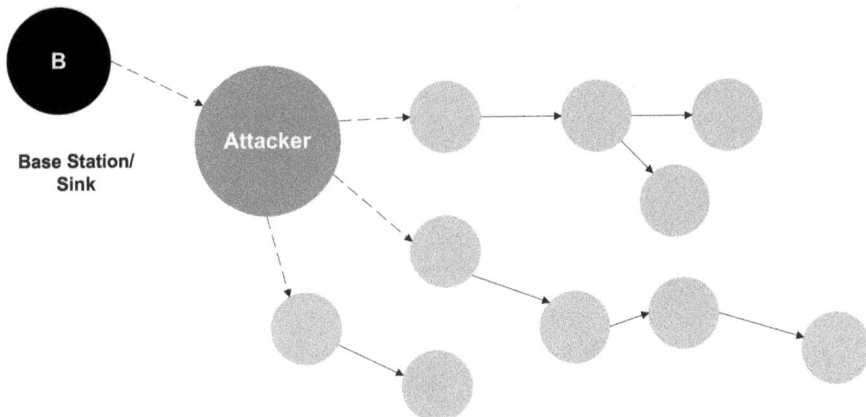

FIGURE 8.2 Blackhole attack.

8.4.1.1.4 Sybil Attack

A security risk in WSN is the Sybil attack, which happens when an attacker uses a rogue node to mimic other nodes in the network. The hacker takes control of the network by establishing a number of false identities, or Sybil nodes, and then manipulates the data and processes therein. A DoS assault, a routing information flood, or both are possible outcomes of these Sybil nodes' malicious actions. Due to the attacker's ability to assume several false identities in a Sybil attack, the WSN's security and functioning are at risk. An attacker can intercept and alter network traffic by manipulating other nodes to route communication through Sybil nodes. Additionally, an attacker can trick genuine nodes into making the wrong routing decisions by manipulating Sybil nodes to provide an inaccurate picture of the network structure, as shown in Figure 8.3.

8.4.1.1.5 Wormhole Attack

In wireless sensor networks (WSNs), an attacker can launch a wormhole attack by stealing packets and rerouting them to another location in the network. To avoid detection by standard network routing algorithms, an attacker can launch this kind of assault by establishing a tunnel, also known as a virtual connection, between two network parts. This gives the hacker access to the network and allows them to breach it by interfering with data transfer, causing data loss, or even introducing harmful material, as shown in Figure 8.4.

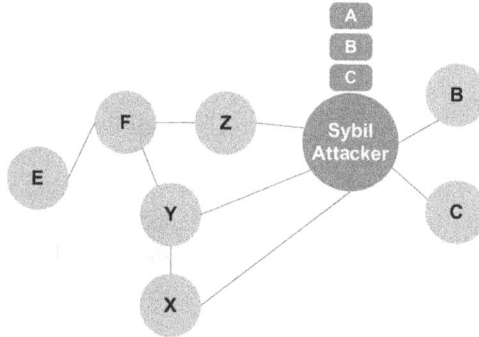

FIGURE 8.3 Sybil attack.

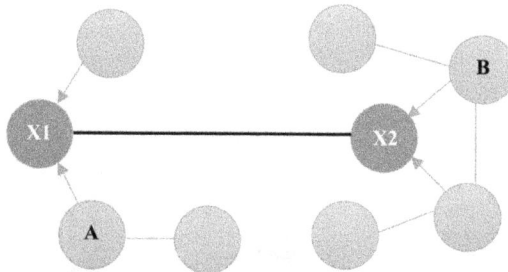

FIGURE 8.4 Wormhole attack.

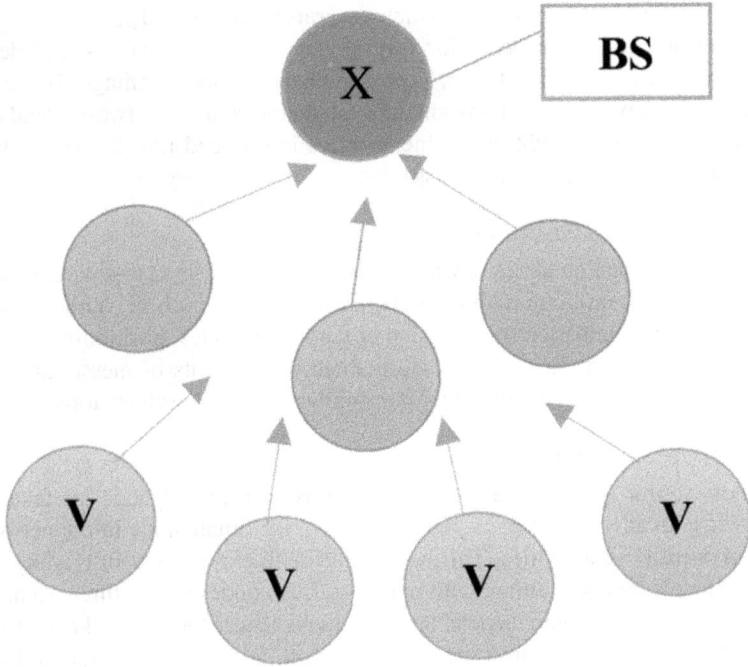

FIGURE 8.5   Hello flood attack.

8.4.1.1.6   Hello Flood

The attacker in a hello flood attack tricks the network's sensors into thinking it's all okay by sending them a barrage of hello packets. An adversary transfers power from one node to another using a routing protocol hello packet. An attacker with extensive processing capabilities and a wide transmission range sends Hello packets to several sensor nodes across the network. As a result, the sensors start to believe the enemy is a nearby neighbour. Consequently, as depicted in the figure, the victim nodes attempt to communicate with the base station (BS) by navigating around the attacker, whom they see as a neighbour, as shown in Figure 8.5.

8.4.1.1.7   Acknowledgement Spoofing

A further attack in WSNs occurs when a malicious actor copies the contents of an acknowledgement message that a node has transmitted and then sends it back to its original sender. An attacker may employ this method to trick the sender into thinking their message reached the destination node when it did not.

8.4.1.1.8   Node Malfunction

Problems with individual nodes seriously threaten the availability and dependability of WSNs. Problems with one node might have a domino effect on the rest of the network, causing data loss or broken connections. In addition, it can potentially create congestion in the network, which can raise latency and lower throughput. The failure of a node can occur for several causes. Possible causes include malfunctioning software or hardware, power outages, or node damage. If the node fails due to these problems, important data can be lost and the network could crash. In addition, attackers might use vulnerabilities introduced into the network by a faulty node to conduct subsequent assaults.

8.4.1.1.9 Collecting Passive Information

A passive eavesdropping attack is when an outsider gains access to a wireless sensor network's data transmission without participating in it themselves. Without interfering with the communication transmission or the sensor nodes' positioning, they may examine the recorded data to learn more about the contents of messages, such as physiological data, location, and other personally identifiable information.

8.4.1.1.10 Artificial Node

In wireless sensor networks, an attack known as an artificial node (or fake node) occurs when an adversarial node pretends to be a legitimate node in the network to carry out destructive activities. This node could act like it has sensor readings, send out false signals, or even tamper with real ones. Such nodes pose a threat to network operations because they can mislead other nodes and cause them to make poor judgements, which in turn can cause the intended job to fail. Authentication and secure communication protocols are two of the many proposed defences to artificial nodes, a major danger to wireless sensor networks.

8.4.1.2 Passive Attacks

An attacker conducting a passive attack just keeps tabs on network traffic to gather sensitive information; they do not change or manipulate the data in any way. Passwords, credit card details, and other sensitive information might be among the data an attacker tries to capture. Several techniques, including packet sniffing, network scanning, and traffic analysis, can be used to execute passive attacks. Intercepting and analysing data packets as they go over a network is known as packet sniffing. An attacker can find possible vulnerabilities in a network by scanning it for devices and services and creating a map of them. To deduce information on communication patterns and transmitted content, traffic analysis examines data flow patterns. Passive attacks are hard to spot since they don't change the data, but they may be stopped by securing the data during transmission with authentication and encryption.

8.4.1.2.1 Congestion

When an attacker deliberately makes the radio frequency (RF) spectrum utilised by the network congested, it can lead to poor network performance or even network failure; this sort of assault is known as a congestion attack on the physical layer. An attacker might launch a congestion attack to disrupt the WBAN nodes' ability to communicate with one another. This interference can be caused by sending out very strong signals on the same channel as the WBAN or by flooding the channel with noise or other signals that lower the SNR and limit the wireless communication's effective range. The assault can cause the sensor nodes to have slower processing speeds, more lost packets, or even a total loss of connectivity, which can affect the WBAN's dependability and performance. As sensor nodes try to retransmit damaged or missing packets, congestion attacks can increase power consumption and shorten battery life.

8.4.1.2.2 Frequency Transmission

To counteract unforeseen congestion or interferences in WSNs, this approach is employed. The process entails altering the broadcast frequency through a sequence, which the receiver must rebuild to recover the initial message. This method may protect the sensor network from harm even in very loud settings. On the other hand, sensor nodes may find the usage of broad-spectrum systems for frequency transmission to be complicated and costly. Despite these obstacles, frequency transmission is a great way to keep wireless sensor networks communicating reliably.

8.4.1.2.3 Frequency Jump

This technique entails quickly adjusting the broadcast frequency to avoid discovery or interfere with communication. This kind of assault usually comes with a hefty price tag and demands a lot of strength. Because of its potential effectiveness in single-frequency networks, it finds widespread application in wireless body area networks. Since many sensor networks can't rapidly adjust to frequency changes, preventing frequency jump attacks isn't always easy. According to Wood, Stankovic, and Son, jammed-area mapping is one way to lessen the impact of frequency jump assaults. Reducing the effects of congestion or interference is the goal of this method, which entails first pinpointing and then mapping those regions. To do this, we may use more sophisticated encryption methods, expand the spectrum of frequencies we use, or employ some other strategy to fortify the network.

8.4.1.2.4 Tapping

The physical layer of WBANs is vulnerable to tapping, a security hazard. Several problems might arise as a result of this assault, which includes illegal access to devices on the network. Due to the vast number of nodes in the network, it might be tough to recognise attackers who may abduct or trap them. The nodes are easy prey for hackers because of their portability and diminutive size. The physical temperature of the devices might be regulated as a possible solution to this problem. If someone tries to tap you, this can assist you in erasing sensitive cryptographic data.

Nevertheless, not all WBANs will benefit from this method, which might be expensive. An alternative approach that has shown promise is using algorithms that mitigate the effect of a single critical component on the network. If all the nodes in a network share a secret key with their agents and neighbours, then compromising only one node will have a localised effect.

8.5 SECURITY PROTOCOLS FOR CONSTRAINED IoT

Due to the significant attack vector potential of smart healthcare sensor networks, it is critical to implement security measures to prevent any assaults. Before, during, and after an assault, the system must be defended. Communication between nodes requires exchanging cryptographic keys for authentication and encryption to provide security services (availability, confidentiality, integrity, and secrecy). On the other hand, everyone knows that encryption systems are the first defence against any assault. It is also important that cryptographic methods identify the most harmful assaults in progress.

Furthermore, these methods need to be compact to make do with the restricted resources of the WSN. To address the issue of resource restriction in sensor devices, classic WSN deployments have suggested many important installation and administration techniques. Because of their minimal resource usage, symmetric cryptography primitives are relied upon by the majority of the suggested methods. As far as sensor nodes are concerned, these systems are the most efficient.

8.5.1 Methods and Protocols Classification

Through a pre-distribution phase, most techniques based on symmetric, asymmetric, or hybrid systems address the key setup problem. The practice of storing encryption keys in memory nodes before deployment is known as pre-distribution in a WSN. Various cryptographic key management techniques have been categorised in the literature.

Key sharing between two or more nodes is the foundation of certain classification algorithms, whereas others depend on using probabilities, combinatorial analysis, etc. We categorise everything into two big families, one for distribution models and the other for key management models. The symmetrical designs are in the second family, whereas the asymmetrical schemes are in the first. Figure 8.4 illustrates the process of classifying these objects. Here, we will go over the most common models mentioned in published papers. Secure communication is essential for Internet of Things nodes, just as for traditional ones. Authentication, secrecy, integrity, and non-repudiation are the main security criteria. Cryptographic primitives, such as signature and verification techniques and encryption and decryption algorithms, form the basis of these security services.

Therefore, a crucial management mechanism is required for these primitives to accommodate IoT devices' limited capabilities and budgetary restrictions, which do not permit the implementation of intricate security systems. In order to function on the constrained resources of nodes, the existing Internet key formation methods are either overly complex or fail to deliver an adequate degree of security. Protocols for establishing keys allow two or more nodes to share a secret, which may then be used as symmetric keys for different types of cryptography. Several security procedures, including those for protecting the authenticity of the source, ensuring data integrity, and maintaining user privacy, rely on symmetric cyphers and message authentication codes to achieve these ends. Schemes that depend on an asymmetric key mechanism and other methods that pre-distribute symmetric keys are the primary types of extant smart sensor security mechanisms and schemes. The categorisation utilised in this publication is illustrated in Figure 8.6.

8.5.2 Symmetric Key Pre-distribution Schemes

The symmetric or secret key approach entails encrypting and decrypting messages using the identical secret key. Key loading into the nodes before deployment is the

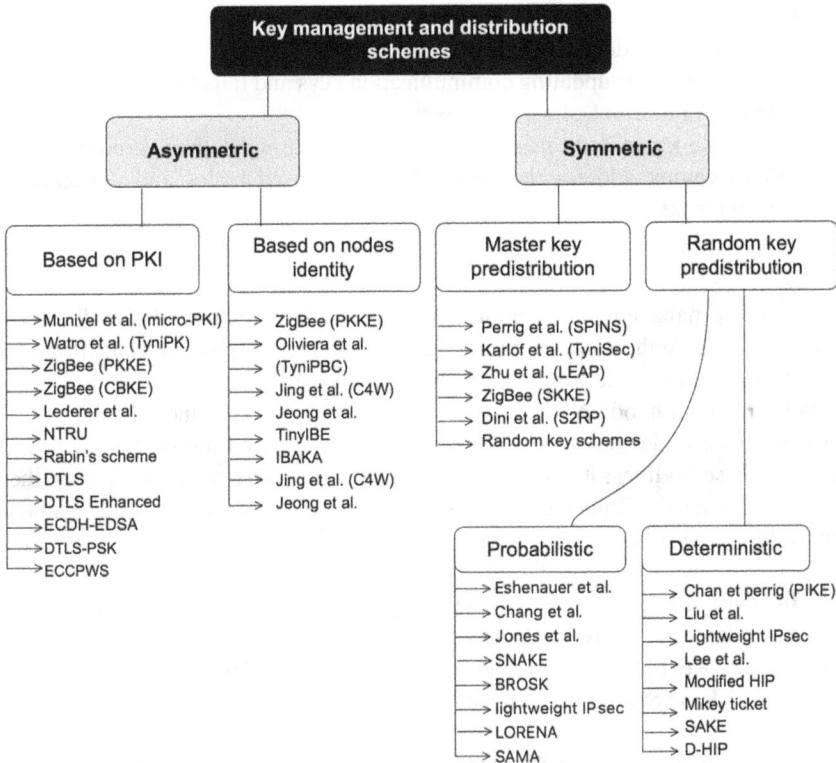

FIGURE 8.6   Key management models for TinyML-enabled devices.

essence of this approach. There are two main types of solutions for key pre-distribution methods in the Internet of Things: deterministic and probabilistic. The primary pre-distribution processes could vary according to what's laid forth here.

8.5.2.1   SPINS

SPINS is a suite of security building blocks proposed by Perig and several other authors (Perrig et al. 2002). A security mechanism that was first suggested for the CWHN. Two protocols, namely µTESLA and the Sensor Network Encryption Protocol (SNEP), form its basis. While only adding eight bytes to each transmission, SNEP guarantees low-cost data secrecy and authentication between two nodes. The enhanced version of TESLA, known as µTESLA, guarantees the broadcast's authenticity. SNEP uses Counter Mode CTR (CounTeR) to implement the RC5 encryption technique. At the outset of the deployment, the network structure only permits communications between the base station(s) and the sensor nodes. With SPINS, a new way is introduced for nodes to extend trust between themselves and the base station to direct connections between nodes. SPINS implements a secure two-party key agreement and an authenticated routing application independently using SNEP and µTESLA, all while consuming little storage, computation, and communication. But there are still some fundamental issues with SPINS, and they are as follows:

- Due to the security routing protocol's paired key pre-distribution strategy, SPINS is overly dependent on the base station.
- SPINS disregards updating communication keys and fails to account for the potential of a denial-of-service attack.
- To achieve forward security, a practical key updating method is required.
- SPINS cannot address the issue of compromised nodes and concealed channel leaks.

8.5.2.2 LEAP

LEAP is a key management system for sensor networks, which stands for "Localised Encryption and Authentication Protocol." Its primary purpose is to facilitate in-network processing while limiting a compromised node's security effect to its local network neighbourhood. An intriguing finding that various kinds of signals sent between sensor nodes have varying security needs inspired the concept of LEAP. Based on these findings, it is clear that a single keying system cannot satisfy these many security needs (Zhu et al. 2003). For each node, LEAP supports the establishment of four types of keys:

- **Individual Key:** Shared with the base station;
- **Pairwise Key:** Shared with another sensor node;
- **Cluster Key:** Shared with multiple neighbouring nodes;
- **Global Key:** Shared by all nodes in the network.

The packets that each node exchanged in a sensor network can be classified into several categories, which are based on different criteria, for example:

- Control packets vs data packets
- Broadcast Packets vs unicast packets
- Queries or commands vs sensor readings and so on.

Depending on its classification, the security requirements for each packet are unique. While secrecy only applies to certain packet types, authentication is required for nearly all types. As stated below, some of the secure communication requirements of sensor networks cannot be met by a single keying technique.

8.5.2.3 LEAP Enhanced

According to LEAP+ (Zhu et al. 2006), one of the most important assumptions is that a node cannot be compromised within Tmin. This theory sounds reasonable; however, it's only feasible under perfect circumstances that Tmin is larger than the one assumed. Using a periodic verification called "Periodic Check" to identify the compromised node is the first of two strategies suggested by Yassine and Ezzati (2016) to overcome this problem, as shown in Figure 8.7. The second model decides whether to remove the shared key after executing a sequence number in each node and comparing it with the information contained in the base station BS. This comparison is done after the pairwise key setup stage.

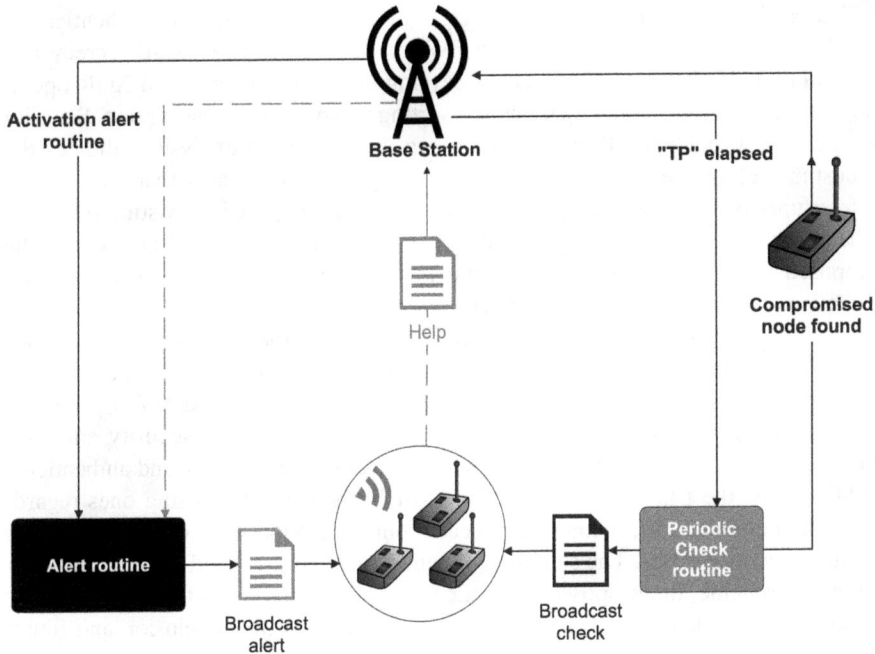

FIGURE 8.7   LEAP enhanced.

8.5.2.4   TinySec

As the initial comprehensive use of a secure design at the data connection layer for WSN, the TinySec Protocol was proposed by Karlof et al.? (Karlof et al. 2004). Two security options are available in this implementation: TinySec-Auth for authentication of messages without data encryption and TinySec-EA for message authentication with data encryption. TinySec employs conventional cryptographic methods, much like SPINS, to provide privacy and message integrity checks. While SPINS uses the RC5 method, TinySec's authors believe that the Skipjack algorithm (Brickell and Davenport 1991) is better suited to WSN. A pre-key computation using 104 bytes of RAM is required by RC5, according to TinySec assessments. Instead of SPINS's CTR encryption mode, TinySec employs CBC, or Cypher Block Chaining. Using the same random values, the CTR will encrypt more packets. Due to their primary purpose in generating encryption key sequences, these numbers' recurrence can compromise the solution's security and make message content discoverable by attackers. Rather than proposing a key distribution technique, TinySec is an implementation tailored to the larger network. For symmetric key sharing across nodes, two keys are required. The first is for message encryption, and the second for message MAC (code) calculation.

Li et al. (2017) proposed a lightweight anonymous mutual authentication and key agreement scheme tailored for centralised two-hop WBANs. The scheme ensures confidentiality, mutual authentication, and anonymity in data transmission, crucial for safeguarding patient information.

Xu et al. (2019) presented a lightweight and anonymous mutual authentication and key agreement scheme tailored for WBANs, prioritising forward secrecy and computational efficiency. Our scheme, leveraging hash functions and XOR operations, ensures forward secrecy without resorting to asymmetric encryption. Security verification through the ProVerif tool, along with informal analysis, validates the robustness of our approach. Comparative analysis demonstrates that our scheme offers superior security and computational efficiency compared to existing solutions. The paper discusses related work, outlines network and threat models, presents the proposed scheme, analyses security and performance, and concludes with future directions in WBAN security enhancement.

Chunka and Banerjee (2021) focused on analysing the security vulnerabilities of a previous scheme by Li et al. (2017), proposing a more efficient authentication and key agreement scheme using cryptographic hash functions and XOR operations. The proposed scheme undergoes rigorous informal and formal security analyses, including BAN logic and ProVerif simulation, to ensure its secrecy and authenticity. Additionally, the paper compares the proposed scheme with existing ones regarding computational cost, memory overhead, communication message exchange, and security functionalities. The subsequent sections delve into related work, a review of Li et al.'s scheme, the proposed authentication and key agreement scheme, security analysis, comparison with existing schemes, and finally, the conclusion and future research directions.

8.5.2.5 Probabilistic Key Distribution

In 2002, Eschenauer and Gligor (2002) presented a system for randomly pre-distributing keys. There are usually three steps to an RKP: distributing pre-keys, discovering shared keys, and establishing path keys. The schema generates a crucial key pool. After that, the sensor nodes receive their keys, chosen from a pool of keys. With some luck, a common commune can share two nodes. When two nodes are not in communication with each other, the third phase has begun. At this point, you can use the secure channel to transition to the key. When key K reaches the opposite node, the procedure terminates. From then on, K and the other node are considered the key pair. This plan proposes several solutions (Chan et al. 2003; Du et al. 2004; Ito et al. 2005). These suggestions focus on enhancing the pre-distribution phase to decrease essential storage space requirements and increase node connections. A pre-distribution system is created by Du et al. (2004) that uses deployment information to prevent key assignments that aren't essential.

A plan based on Du et al. (2004) applied to two-dimensional locations is developed by Ito et al. (2005). They provide an improved key connectivity price-density function. Chan et al.'s (2003) paper has also been translated into French for the convenience of French speakers. Node A discovers every potential link to node B. This is the fundamental concept. These arbitrary numbers serve to secure the shared keys for A and B. If you don't want to be able to spy on every path between them, the generous key will be shared by both nodes. Establishing the session key between all nodes is not guaranteed by the probabilistic key distribution, even with the path key's setup phase. There is a possibility that no common key exists between any two languages.

8.5.2.6 Deterministic Key Distribution

The key schemes discussed in this section use a predetermined procedure to establish the key pool and disperse the whole network. Including an unbiased third party during key boot is a defining feature of deterministic solutions' key schemas.

The ease of the offline key distribution method makes it a popular choice in WSN. Every node could share a pair of keys, depending on the protocol. The session is subsequently created after the third party has arrived. Since it does not need costly cryptographic computations like asymmetric methods, offline key distribution uses less power. The sensitive information kept in a sensor node can be compromised in the event of a physical assault on the node. That means the criminal can potentially compromise the whole network or even several nodes that share the same secret key as the one they're investigating. The model for secure key exchanges between sensor nodes has been developed in several previous publications using mathematical principles. Even in the Internet of Things setting, these plans have their uses. Bivariate polynomials provide the basis of the most well-known schemes (Fanian et al. 2010; Liu et al. 2005). A bivariate n-polynomial degree $f(x,y)$ is assigned to a common node A in these systems. The value of $f(IdA, IdB)$, where IdA and IdB are the identities of A and B, respectively, may be derived. Since $f(IdA, IdB)$ is equivalent to $f(IdB, IdA)$, B may obtain the same key pair.

One such approach, the Bloom scheme (Blom 1984), uses the secret key that nodes A and B share to create a secret symmetric matrix D. They provide a public matrix IA for A and an IB for B, respectively. For A, the private key is privA = DxIA; for B, it is privB = DxIB. The last step is determining the key pair by solving either (privA × IB) or (privB × IA). Both of these cases involve the dilemma of the unchanging. Two key settlement methods that include key management are SNAKE (Seys & Preneel 2002) and BROSK (Lai et al. 2002). All nodes on the same network are assumed to share a primary secret key by both protocols. Each communicating node in SNAKE uses the pre-shared key to produce two random nonces, forming the session together. The nuncio, an important message in negotiations, is aired by BROSK. After a node has received messages from its neighbours, it may construct the session key by determining the MAC of two nuncios.

According to Raza et al. (2011), an IP-based WSN may be secured using the industry-standard IPsec protocol by using 6LoWPAN. Their proposed methods manage packet sizes while integrating IPsec with the 6LoWPAN layer and compressing the AH and ESP headers. Despite their usefulness for origin authentication, message integrity, and IP packet privacy, the AH and ESP techniques aren't equipped to deal with key exchange. The use of a pre-shared key facilitates the manual establishment of security relationships.

8.5.2.7 LORENA

Coelho et al. (2022) addressed the security challenges in the Internet of Health Things (IoHT), particularly in wearable medical devices and remote health monitoring systems. With the increasing proliferation of IoT devices in healthcare, ensuring the secure transmission of sensitive patient data becomes paramount. The paper proposes a novel protocol called LORENA (Low memORy symmEtric-key geNerAtion method based on groups) to generate symmetric keys with low resource consumption using physiological signals, particularly electrocardiogram (ECG) data. LORENA establishes secure communication channels and facilitates key device agreement, ensuring data confidentiality and integrity. LORENA achieves efficient key generation and transmission by leveraging the human body as a communication medium and employing approximate computing techniques while minimising resource usage. The paper's contributions include defining communication protocols, evaluating key generation efficiency, and providing a scalable solution suitable for real-world deployment with low-cost microcontrollers.

8.5.3   ASYMMETRIC KEY SCHEMES

By using a "public" key for encryption and a "private" key for decryption and signing, the asymmetric or public key approach ensures that messages remain secure. According to Nguyen et al. (2015), there are primarily two types of asymmetric schemes: public key encryption for key transmission and classic asymmetric approaches for key management. The following sections will concisely analyse several asymmetric key schemes that can be used in the Internet of Things.

8.5.3.1   Key Transport Based on Public Key Encryption

This subcategory examines the key establishment schemes in which the public key is used to carry secret data or to negotiate a session key. Several methods are used to generate the public and private key pair. This subcategory classifies these mechanisms according to public/private key generation methods. Figure 8.7 gives an example of a communication scenario between two entities, A and B. In this scenario, A and B can use the public keys to create an encrypted channel. The Certificate Authority (CA) can participate in verifying the identity of the message sender when certificates are supported. This method can be expensive for resource-constrained sensor nodes, especially when using a traditional algorithm such as RSA. Without a verifiable relationship between the public key and identity (i.e., cryptography based on identity, cryptographic identification, or CA mediation), this approach becomes vulnerable to man-in-the-middle attack. Indeed, A and B cannot authenticate the identity of the other. When communicating with B, an attacker can generate any public/private key and pretend to be A.

8.5.3.2   Micro-PKI

A more condensed form of traditional PKI, micro-PKI (Public Key Infrastructure Micro) is the approach that Munivel and Ajit (2010) suggest for WSN. A public key and a private key are stored in the base station. Nodes in the network authenticate the base station using the public key, while the base station uses the private key to decode data received by the nodes. Before deployment, every node stores the base station's public key. Two different forms of authentication (Handshake) are incorporated into the authors' approach. Nodes in a network initially authenticate with one another and the base station. The node creates a symmetric session key using the base station's public key for encryption. The authors suggest including a message authentication code (MAC) that uses the same encryption key as the message to guarantee the authenticity of messages sent and received. To facilitate the addition of new nodes to the network, the public key of the base station is saved in these nodes before deployment.

8.5.3.3 TinyPK

Using public keys and the Diffie-Hellman principle, Watro et al. (2004) provided TinyPK, a technique for establishing a secret key between two WSN nodes. Nodes' public keys are signed by a trustworthy authority in TinyPK. Each node receives the CA key before deployment so that they may verify each other's key neighbours. Nodes use many resources due to the RSA algorithm's selection for encryption. So, even the most fundamental tasks might take a few seconds, which affects responsiveness and shortens the network's lifespan.

8.5.3.4 PKKE & CBKE

Key establishment in Zigbee's PKKE and CBKE protocols relies on node identities. Using these IDs, we can generate a unique key that can be used to unlock any two nodes in a network. Still, the two nodes need to communicate to generate the shared key. So, to generate a key, procedures necessitate the exchange of several messages in both directions. Several solutions have been suggested to eliminate these interactions, which would benefit the intermediate and power nodes that wish to keep a secret. The acronym for "Identity-Based Non-Interactive Key Distribution Scheme" is "ID-NIKDS" in the cryptography community (Steinwandt & Suárez 2011).

8.5.3.5 C4W

A novel approach, C4W, was put out by Jing et al. (2006), and it relies on the nodes' identities to determine public keys. By using their own identities, nodes may deduce the public keys of other nodes. In what ways may a certificate be superseded? Keys (private/public key ECC) and public information about the network nodes are placed into the base station and the nodes before deployment. Without the need for certificates, using the Diffie-Hellman key exchange principle, the C4W technique generates a single shared key between any two nodes.

8.5.3.6 pDCS (Privacy-Enhanced Data-Centric Sensor Network)

A privacy-enhanced data-centric sensor networks (pDCS) architecture, "data-centric sensor networks, enhanced privacy" (Shao et al. 2009), uses rectangular cells divided by Steiner's Euclidean trees to organise the network. Within each cell, there are sensors and cell keys. Mobile data "sinks" collect and transmit this information, but the encryption prevents an unauthorised party from accessing the original sensor that detected an event. Once again, a Bloom filter reduces the volume of control data generated by pDCS. ERP-DCS, "an efficient protocol for key regeneration for DCS networks" (Shun et al. 2013), has been developed to enhance the key management mechanism when an agent is compromised and recognised. An exclusion system known as EBS has been employed (Exclusion Basis System).

8.5.3.7 Zigbee

While Zigbee's (2006) focus is more general, it uses the IEEE 802.15.4 stack, which is utilised in sensor networks on occasion. As things stand, he has the potential to become the de facto norm for all "Internet of Things" linked devices. When coupled with sensors, it safeguards data by encrypting it, authenticating its user, and preventing replay attacks. With the introduction of the concept of a "trust centre" by Zigbee, key management is centralised. Although alternative designs offer additional security, this one comes at the expense of headers and computations.

Using out-of-band communications or pre-distribution of the public key are two examples of how such processes work. These methods allow for a limited amount of message exchanges, but they can't handle large networks since every device has to know everyone else's public key. Some "raw public key encryption" methods have been suggested for WSN networks, including NtruEncrypt (Gaubatz et al. 2005) and Rabin (1978). RSA, an algorithm in many cryptosystems, is likewise based on the difficulty of the factorisation issue; Rabin's schema is quite close to it. Interestingly, the system's power consumption for decryption processes is the same as RSA, while maintaining the same level of security. Since only one equation is required to encrypt a message, it offers a considerably quicker technique for encryption processes. One of the alternative cryptographic systems is NtruEncrypt, which uses a trellis with RSA and ECC primitives. Smart cards and RFID tags, which have limited resources, are ideal for this technique because of its efficiency. The Rabin method, NtruEncrypt, and ECC are the three suggested PKC techniques for limited devices that are compared in Gaubatz et al. (2005). According to the findings, NtruEncrypt is the most energy-efficient operation on average.

Nevertheless, this encryption scheme may cause packet fragmentation at lower levels and several retransmissions when communication faults occur, frequently requiring huge messages. If the transmitting power is the most significant and limiting component, protocols based on "raw public key encryption" have a modest message exchange need, which is highly favourable. An authentication approach for the Internet of Things (IoT) based on Two-level Session Keys (TSKs) was suggested by Mahmood et al. (2017). Additionally, a technique for associating nodes is presented. End-to-end users may communicate securely and with minimal overhead using TSK.

Many security service standards established by the Internet Engineering Task Force (IETF) include TLS as their preferred protocol (Turner & Polk 2011). Nevertheless, TLS is not recommended for optimal security in the IoT, as stated in Kothmayr et al. (2012) and Raza et al. (2013). TLS is typically used with dependable transport protocols like TCP, which aren't ideal for devices with limited resources because of their congestion management method. Datagram Transport Layer Security (DTLS) is a newer proposal to replace Transport Layer Security (TLS) in very limited settings. It offers the same robust security as TLS but operates on the unstable UDP transport technology. Using a certificate is, at its core, costly. The researchers examined the following software and hardware enhancements to lower energy consumption: Implementation of hardware accelerators for cryptography: The computations involved in cryptography are handled by hardware accelerators. A technique for DTLS implementation utilising sensor node hardware was proposed by Kothmayr et al. (2012). This method is based on the premise that all sensors have a TPM (Trusted Platform Module) installed. Hardware support for the RSA algorithm and tamper-proof key generation and storage is provided by an integrated device known as a TPM. Before deployment, a publisher with a trusted CA certificate and a trusted publisher certificate integrated with a trusted hardware module (TPM) must be stored on the publisher. We offer authentication using the DTLS pre-shared key encryption key for publishers without TPM chips. This key requires minimal randomly generated bytes to be placed onto publishers before deployment. For the AC server to provide the device keys to those with the proper authority, this secret must also be accessible to them. In addition to a high degree of security in trust building with the aid of an authorised third party, this solution offers inexpensive energy, end-to-end latency, and overhead memory, as well as message integrity, secrecy, and authenticity.

Recent research by Maleh et al. (2016) aim to mitigate denial-of-service attacks by lowering the communication cost of the DTLS protocol and strengthening the vulnerability of cookie exchange during connection establishment. To lessen the burden on the network and save space, the Constrained Application Protocol (CoAP) incorporates the upgraded DTLS protocol.

Shamir (1984) created the initial version of identity-based cryptography. An individual's or group's public key is defined by this cryptographic method as a known string (identity). A third party, known as a PKG (Public Key Generator), uses each entity's public key to create its private key, as seen in Figure 8.8. This technique is especially beneficial for WSNs since it eliminates certificate requirements. To set up secure communication utilising their identities, any sensor node may easily produce the public key of any other node. The verification of the legitimate sensor identification also helps the revocation procedure. Since PKG has access to the private keys of every node in the network, ID-based schemas are susceptible to key deposit attacks.

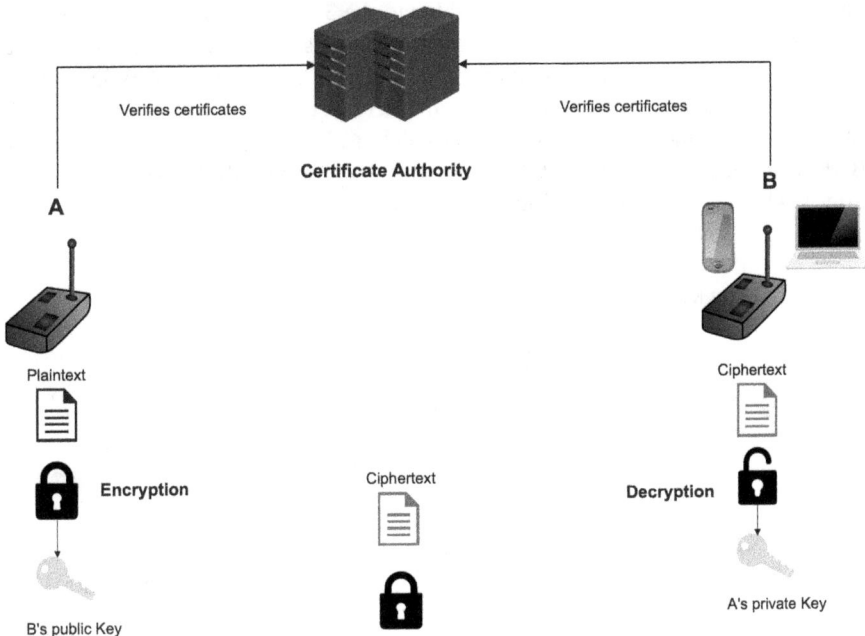

FIGURE 8.8 Public key transport mechanism.

According to Yang et al. (2013) and Granjal et al. (2013b), the ECC primitive is the most common way to use the IBE paradigm in a limited setting. Other primitives have their implementations; for instance, ElGamal-type RSA or IBE. But, because there are a lot of exponentiation operations, and each one has a big exponent, they are costly for the limited nodes. Using the design from Boneh and Franklin (2003) as inspiration, IBAKA is an IBE system proposed by Yang et al. (2013). To create a session key, though, they modify the IBE technique so it can work with an ECDH key exchange. Every time a secret key is started, their proposal calls for two bilinear pairings and three multiplications of scalar points.

8.5.3.8 Asymmetrical Schemes for Key Agreement

Protocols for key agreement based on asymmetric primitives in the Internet of Things fall under this class. According to many studies, a key agreement protocol is a way for two or more parties to develop a shared secret without anybody else being able to guess how much that secret is worth (Nguyen, Laurent, and Oualha 2015). An example of an asymmetric key agreement is shown in Figure 8.9. A notable example of a symmetric key chord is the Diffie-Hellman (DH) protocol and its variations (E. Rescorla 1999). Nevertheless, based on how nodes are classified in lwig terminology according to their resource capacity, Diffie-Hellman protocols are deemed inappropriate and costly for limited nodes, namely those in classes 0 and 1 (Bormann et al. 2013). Several variants of the Diffie-Hellman protocol are contemplated in limited settings, employing ECC (ECDH). Compared to RSA, the key size of the ECDH cryptographic primitive is much smaller. The US National Institute for Standards and Technology (NIST) has demonstrated that a 256-bit key, generated via an elliptic curve, is preferable to the 3,072 bits used by RSA and the DH protocol in order to attain the 128-bit AES key's security level. For example, a framework that enables end-to-end adaptive security in the context of WSNs bound for the Internet and end-to-end address transport layer security with delegated ECC public key authentication is provided (Granjal et al. 2013a). For sensor networks, IBAKA provides a hybrid of ECDH and BIE (Yang et al. 2013). Using the Boneh pattern, which is based on identification, the system ensures that message exchanges

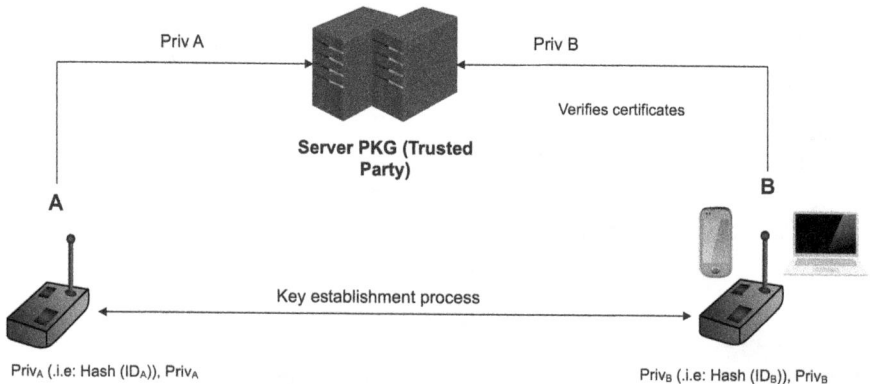

FIGURE 8.9 Identity-based cryptography scheme.

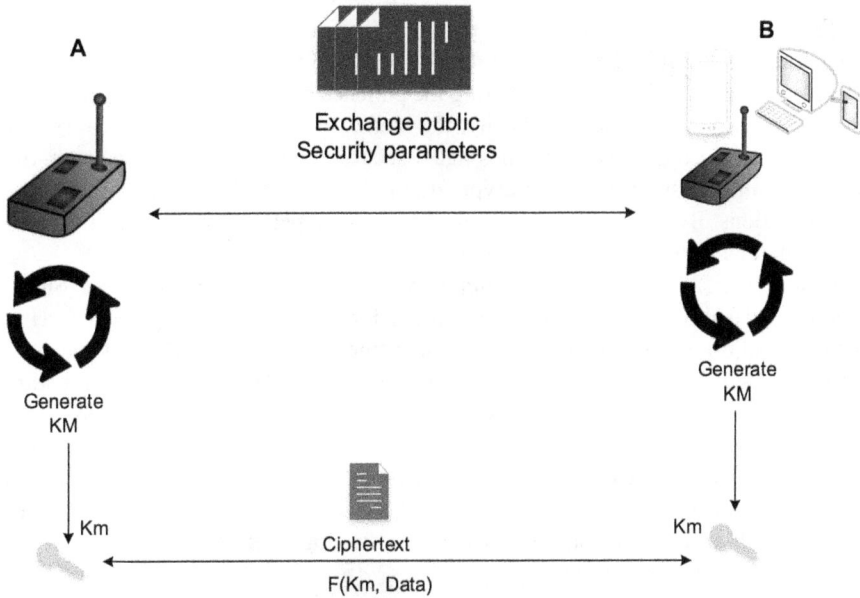

FIGURE 8.10   Key agreement based on asymmetric mechanisms.

remain secret, building on the foundation of the ECDH protocol (Boneh & Franklin 2003) (Figure 8.10).

Using public key methods and the Rabin scheme, Hayajneh et al. (2014) presented a lightweight authentication protocol for WBAN, which is well-suited to small devices with limited resources. In a medical setting, the technology is intended to be implanted in the bodies of individuals suffering from various ailments. Actuators, a physician, sensors, and a coordinator of nodes are the four components of the system. In the system, these nodes securely communicate with one another.

Building on the work of Lu et al. (2015), the study of Chaudhry et al. (2015) introduces an improved protocol used for authentication in Telecare Medic Information Systems (TMIS). Patients' privacy, their identification, and the TMIS server were all determined to be at risk under the initial protocol, according to the authors. To fix these flaws, they came up with a new protocol that is more secure but also more computationally expensive. To guarantee the protocol's resilience against assaults, the ProVerif tool was used for testing.

In Abdmeziem and Tandjaoui (2015), the authors present a secure key management strategy that e-health apps may use to keep patient data private, intact, and readily available. Key distribution and administration are made easier by the protocol's hybrid method, which mixes symmetric and asymmetric encryption. Key management and setup between the sensors and the server are handled by a trusted third-party authority (TPA) in the protocol. The authors also suggested a safe approach for establishing session keys that employs a one-time password for both the server and sensor to authenticate each other. Results from a simulation test of the proposed protocol demonstrate its efficacy, security, and ability to manage the essential process for e-health applications.

Telecare Medic Information Systems utilised a newly introduced protocol (Chaudhry et al. 2015). The paper suggests a method for authenticating a healthcare server in a telecare HIS while protecting the patient's privacy. The protocol has four parts: the patient, the trusted authority, the medical server, and the medical sensor node. To ensure the transmitted messages' security, authenticity, and privacy, the protocol employs symmetric key cryptography, message authentication codes, and hash functions. By using a pseudonym rather than the patient's real name, the technique guarantees the patient's privacy.

Feng et al. (2020) proposed a collaborative authentication protocol for Smart Electronic Health Record (SEHR) systems, addressing the challenge of ensuring security and privacy in healthcare data management. The paper highlights the need for robust authentication mechanisms in SEHR systems due to the sensitive nature of patient data and the potential risks associated with unauthorised access. By introducing a collaborative authentication protocol based on two-party computation, the paper aims to enhance efficiency and security while ensuring fairness between patients and doctors. Additionally, the paper thoroughly analyses the proposed protocol's provable security and performance, demonstrating its effectiveness for practical implementation in SEHR systems.

A novel privacy-preserving mutual authentication system for WBANs was proposed by Jegadeesan et al. (2020), who also introduced a privacy-preserving protocol. The protocol aims to provide private and efficient data transfer between a user's wearable sensors and an off-site server. The three main components of the suggested protocol are the user, the proxy node, and the distant server. The protocol protects the user's anonymity, which permits the user to remain anonymous to both the proxy node and the distant server. Its unique encryption and decryption technique, which is based on elliptic curve cryptography and hash functions, achieves secure communication between entities. The four primary components of the suggested protocol are signing up, verifying identity, creating a session key, and exchanging messages. The user's authentication and session key generation occur during the registration step. During the authentication phase, the session key is used for mutual authentication between the user and the proxy node.

Additionally, the proxy node verifies the distant server. The user and the remote server establish the session key during the session key establishment phase. The user and the remote server have created a session key and are ready to exchange encrypted communications. WBANs can benefit from the suggested EPAW protocol's privacy-preserving mutual authentication approach, which is both efficient and safe. Connecting the user's wearable sensors to a distant server safely, the protocol also protects the user's anonymity. Despite being able to operate in WBANs with limited resources, the simulation-based trials demonstrate that the suggested protocol has little computing cost and overhead.

It was suggested in 2020 that wireless sensor networks use a new authentication and key agreement system (Rehman et al. 2020). User nodes (UNs), gateway nodes (GNs), and remote healthcare servers (RHSs) are the three components of the suggested design. The scheme's use of hash functions and symmetric key encryption ensures secure communication between entities. First, the UN and GN create a session key; second, the GN and RHS use the key established during the first session to share the key. This is the first of two steps in the key agreement procedure. To ensure that communications sent between entities are legitimate, the authentication procedure uses message authentication codes (MACs). The suggested approach safeguards data from attacks like replay, impersonation, and man-in-the-middle while simultaneously facilitating mutual authentication between entities. To ensure the scheme is secure and works as intended, we ran it using the AVISPA tool. The security study shows that the suggested scheme is safe and efficient; hence, it may be used in WBANs. Currently, existing techniques use more memory, take more calculation time, and have more communication overhead than the suggested approach.

Olufemi and Adedamola (2020) proposed a Secure Addressing and Mutual Authentication (SAMA) protocol. With the rapid integration of healthcare and IoT technologies, ensuring the secure communication between patients and doctors remotely has become increasingly vital. The paper emphasises the need for a robust protocol to uniquely identify smart medical devices (SMDs) and establish mutual trust between these devices and the medical authentication server (HS). The SAMA scheme aims to achieve this by utilising a modified standard IPv6 address format for secure addressing, incorporating password authentication, unique identity verification of doctors and SMDs, and establishing secure communication channels with session keys. By encapsulating each packet without increasing packet size, the proposed protocol aims to mitigate various security threats and ensure patient data's integrity and confidentiality.

Pirmoradian et al. (2023) evaluated the security of a recent authentication scheme proposed by Sowjanya et al. (2020) for WHMS, based on elliptic curve cryptography (ECC). Our analysis reveals vulnerabilities to passive insider secret disclosure and replay attacks in their scheme. To address these weaknesses, we propose a novel ECC-based authentication scheme, ECCPWS, designed to mitigate these vulnerabilities effectively. We provide informal and formal security proofs for ECCPWS, including analysis using the Real-Or-Random (ROR) model, Burrows-Abadi-Needham (BAN) logic, and verification using Scyther and AVISPA tools. Our findings demonstrate that ECCPWS offers comprehensive security against various security threats, ensuring the integrity and privacy of patient data in WBANs.

Joha et al. (2024) presented a secure and intelligent Industrial Internet of Things (IIoT) framework that integrates AI-driven real-time short-term active and reactive load forecasting with anomaly detection in a real-world industrial energy management context. It introduces a hybrid deep learning model—TCN-GRU-Attention—for accurate one-day-ahead forecasting of both active and reactive power loads, achieving superior performance over conventional models. In parallel, the study proposes an optimised Isolation Forest algorithm for real-time anomaly detection, effectively identifying irregularities even under transient conditions of industrial appliances. Designed for deployment on resource-constrained edge devices (e.g., Jetson Nano), the system ensures low-latency operation and computational efficiency. To secure data transmission and system integrity, the framework employs TLS/SSL protocols and encrypted credentials. Overall, this research contributes a scalable, energy-efficient, and secure IIoT architecture that supports reliable energy optimisation and predictive analytics for industrial environments.

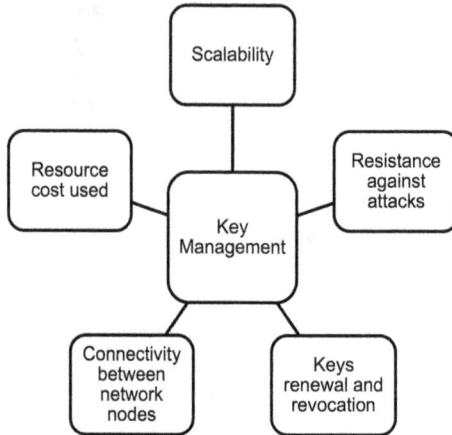

FIGURE 8.11   The criteria for comparison of methods of key management.

8.6   ANALYSES AND COMPARISON

8.6.1   ANALYSES METHOD

To compare the various key management approaches, a number of characteristics are considered. Figure 8.11 displays the primary criteria. We start by limiting the resources that nodes have. Now that the nodes have been installed to gather the data, the suggested key management mechanism must consider that. They rely on their embedded energy and memory space for data storage and application role assurance.

Additionally, the solution needs to be scalable, dynamic, and flexible. Resistance to assaults is another need that must be met. If an adversary manages to capture a node and stores information about it, they can use that data to launch more attacks or otherwise control the network. Before dispersing keys, the key management system should be able to identify compromised nodes and verify their authenticity. Key renewal and revocation are the last requirements. It can be considered just as crucial as key distribution. Keys that have expired or been detected by an adversary must be revoked. It is also necessary to refresh the keys for the secure links regularly. A network's connection ensures nodes have more secure pathways to transmit data. Key distribution methods should be able to guarantee reliable network connectivity. If a node leaves or is captured, it might affect the ability of other nodes to connect to the network. This aspect has to be considered by the distribution technique when suggesting new safe pathways.

8.6.2   COMPARISON AND DISCUSSION

We have analysed various key management schemes for TinyML-enabled devices and categorised their diagrams in Figure 8.6. Table 8.1 shows the results of comparing these diagrams using the same criteria as shown in Figure 8.11. Be aware that the table's evaluation of memory storage just considers the amount of keys stored in the nodes, ignoring the size of code algorithms and cryptographic primitives (Table 8.3).

TABLE 8.3

Key Management Techniques for WSN as Suggested and Compared

| | | | | | Criteria of Comparison | | | | | | |
| | | | | | Resistance Against Attacks | | | | Resource Cost Used | | |
Type	Authors	Based on	Scalability	Connectivity	Information Collection	Communication Perturbation	Data Aggregation and Resource Exhausted	Capture of Physical Nodes	Memory (Key Store)	Calculation and Energy Consumption	Renewal and Revocation
Symmetric key schemes	SNAKE (Seys, & Preneel 2002)	Deter.	–	+		–		–	+	–	–
	BROSK (Lai et al. 2002)	Deter.	–	+		–		–	–	–	–
	Chan et al. (2003)	Proba.	–	+		–		+	–	+	–
	Chan and Perrig (2005)	Deter.	–	+		–		+	+	–	–
	Perrig et al. (2002)	MK + BS	–	+		–		–	+	+	+
	Zhu et al. (2003)	MK	+	+		–		–	+	–	+
	Lightweight IPsec (Raza et al. 2011)	Deter.	+	+		–		–	+	+	–
	DTLS-PSK (Granjal et al. 2013b)	PKI	+	+		–		–	+	–	+
	Yassine and Ezzati (2016)	MK + BS	+	+		–		+	+	–	–
	LORENA (Coelho et al. 2022)	Proba.	+	+		–		+	+	–	+

(Continued)

TABLE 8.3 (Continued)
Key Management Techniques for WSN as Suggested and Compared

Schemes					Criteria of Comparison						
					Resistance Against Attacks				Resource Cost Used		
Type	Authors	Based on	Scalability	Connectivity	Information Collection	Communication Perturbation	Data Aggregation and Resource Exhausted	Capture of Physical Nodes	Memory (Key Store)	Calculation and Energy Consumption	Renewal and Revocation
Public key schemes	Zigbee (2006)	ID	+	-		+		+	-	-	+
	Munivel and Ajit (2010)	PKI	-	-		+		-	+	+	-
	Watro et al. (2004)	PKI	-	-		-		+	-	-	+
	DTLS modified (Raza et al. 2013)	PKI	+	+		-		+	+	-	-
	IBAKA (Yang et al. 2013)	Node identity	+	-		-		+	-	-	+
	DTLS Enhanced (Maleh et al. 2016)	PKI	+	+		+		+	-	+	-
	Feng et al. (2019)	PKI	+	+		-		+	-	-	+
	EPAW (Jegadeesan et al. 2020)	PKI	+	+		+		-	+	+	-
	SAMA (Olufemi & Adedamola 2020)	Proab.	+	+		+		+	-	+	+
	ECCPWS (Pirmoradian et al. 2023)	PKI	+	+		+		-		-	+
	Joha et al. (2024)	AI + TLS/SSL									-

Memory, Connectivity, Resilience, Computational Complexity, Communication Complexity, and Renewal and revocation are the five evaluation metrics. These metrics have two possible values, representing the degree to which a particular protocol supports a property: + (good or medium performance level) and – (poor performance level).

8.7 CONCLUSION

Based on the findings of this research, several important recommendations emerge for different stakeholders involved in TinyML deployments. Adopting multi-layered security approaches is critical to effectively addressing TinyML vulnerabilities by implementing security mechanisms across all levels of the system architecture, from hardware components to applications. Optimisation techniques such as model quantisation and distillation should be employed to enable the integration of robust security features within the tight resource constraints of TinyML devices. Organisations are encouraged to develop tailored security policies that recognise the unique challenges posed by TinyML environments and ensure that security practices align with operational realities. Furthermore, investing in staff training and cultivating specialised TinyML security skills is crucial for the success of implementations. The research community should focus on creating security solutions specifically adapted to the constraints of TinyML while also establishing standardised benchmarks and evaluation methodologies to facilitate objective comparisons and improve solution effectiveness.

This analysis primarily concentrated on the technical aspects of TinyML security, with limited attention given to economic and social factors. Future research could explore the economic impacts of security measures as well as the broader social implications of widespread TinyML deployment. Additionally, the evaluation of security solutions in this study was mainly based on theoretical analysis and limited case studies. More extensive empirical evaluations through real-world deployments are necessary to validate and refine these recommendations. The integration of emerging technologies such as quantum computing and explainable artificial intelligence into TinyML security strategies presents promising opportunities for enhancing protection mechanisms. Finally, developing adaptive security frameworks capable of evolving with changing threat landscapes represents a vital direction for future research.

REFERENCES

Abdmeziem, M. R., and D. Tandjaoui. 2015. "An End-to-End Secure Key Management Protocol for e-Health Applications." *Computers & Electrical Engineering* 44: 184–97.

Abouzakhar, N. S., A. Jones, and O. Angelopoulou. 2017. "Internet of Things Security: A Review of Risks and Threats to Healthcare Sector." In *2017 IEEE International Conference on Internet of Things (I Things), and IEEE Green Computing and Communications (GreenCom), and IEEE Cyber, Physical and Social Computing (CPSCom), and IEEE Smart Data (SmartData),* Exeter, UK, IEEE, 373–78.

Boneh, D., & M. Franklin. 2003. "Identity-Based Encryption from the Weil Pairing." *SIAM Journal on Computing* 32(3): 586–615.

Bormann, C., M. Ersue, and A. Keranen. 2013. "Terminology for Constrained Node Networks." *Draft-Internet.*

Butt, S. A., J. L. Diaz-Martinez, T. Jamal, A. Ali, E. De-La-Hoz-Franco, and M. Shoaib. 2019. "IoT Smart Health Security Threats." In *2019, the 19th International Conference on Computational Science and Its Applications (ICCSA),* St. Petersburg, Russia, IEEE, 26–31.

Chan, H., and A. Perrig. 2005. "PIKE: Peer Intermediaries for Key Establishment in Sensor Networks." *Ieee Infocom* 1(Cc): 524–35. doi:10.1109/INFCOM.2005.1497920.

Chan, H., A. Perrig, and D. Song. 2003. "Random Key Predistribution Schemes for Sensor Networks." *Proceedings—IEEE Symposium on Security and Privacy*, Berkeley, CA, USA, 2003-Janua: 197–213. doi:10.1109/SECPRI.2003.1199337.

Chaudhry, S. A., K. Mahmood, H. Naqvi, and M. K. Khan. 2015. "An Improved and Secure Biometric Authentication Scheme for Telecare Medicine Information Systems Based on Elliptic Curve Cryptography." *Journal of Medical Systems* 39: 1–12.

Chunka, C., and S. Banerjee. 2021. "An Efficient Mutual Authentication and Symmetric Key Agreement Scheme for Wireless Body Area Network." *Arabian Journal for Science and Engineering* 46(9): 8457–73. doi:10.1007/s13369-021-05532-8.

Coelho, K. K., M. Nogueira, M. C. Marim, E. F. Silva, A. B. Vieira, and J. A. M. Nacif. 2022. "LORENA: Low MemORy SymmEtric-Key GeNerAtion Method Based on Group Cryptography Protocol Applied to the Internet of Healthcare Things." *IEEE Access* 10: 12564–79. doi:10.1109/ACCESS.2022.3143210.

Du, W., S. Chen, W. Du, J. Deng, Y. S. Han, S. Chen, and P. K. Varshney. 2004. "A Key Management Scheme for Wireless Sensor Networks Using Deployment Knowledge." In *Electrical Engineering and Computer Science*, Hong Kong.

Eschenauer, L., and V. D. Gligor. 2002. "A Key-Management Scheme for Distributed Sensor Networks."

Fanian, A., M. Berenjkoub, H. Saidi, and T. A. Gulliver. 2010. "A Scalable and Efficient Key Establishment Protocol for Wireless Sensor Networks." In *2010 IEEE Globecom Workshops*, GC'10, Miami, FL, USA, 1533–38. doi:10.1109/GLOCOMW.2010.5700195.

Feng, Q., D. He, H. Wang, L. Zhou, and K.-K. R. Choo. 2020. "Lightweight Collaborative Authentication with Key Protection for Smart Electronic Health Record System." *IEEE Sensors Journal* 20(4): 2181–96. doi:10.1109/JSEN.2019.2949717.

Gaubatz, G., J. P. Kaps, and E. Ozturk. 2005. "State of the Art in Ultra-Low Power Public Key Cryptography for Wireless Sensor Networks." In *Proceedings of the Third IEEE International Conference on Pervasive Computing and Communications*, Kauai, HI, USA, 146–50. doi:10.1109/PERCOMW.2005.76.

Granjal, J., E. Monteiro, and J. Sa Silva. 2013a. "A Framework towards Adaptable and Delegated End-to-End Transport-Layer Security for Internet-Integrated Wireless Sensor Networks." In the *2nd Joint ERCIM EMobility and MobiSense Workshop*, St Petersburg, Russia, 34.

Granjal, J., E. Monteiro, and J. Sa Silva. 2013b. "End-to-End Transport-Layer Security for Internet-Integrated Sensing Applications with Mutual and Delegated ECC Public-Key Authentication." *IFIP Networking Conference*, Brooklyn, NY, USA, 2013: 1–9.

Hayajneh, T., A. V. Vasilakos, G. Almashaqbeh, B. J. Mohd, M. A. Imran, M. Z. Shakir, and K. A. Qaraqe. 2014. "Public-Key Authentication for Cloud-Based WBANs." In *Proceedings of the 9th International Conference on Body Area Networks*, London, UK: 286–92.

Ito, Takashi, Hidenori Ohta, Nori Matsuda, and Takeshi Yoneda. 2005. "A Key Pre-Distribution Scheme for Secure Sensor Networks Using Probability Density Function of Node Deployment." In *Proceedings of the 3rd ACM Workshop on Security of Ad Hoc and Sensor Networks—SASN '05*, Alexandria VA, USA: 69. doi:10.1145/1102219.1102233.

Jegadeesan, S., M. Azees, N. Ramesh Babu, U. Subramaniam, and J. D. Almakhles. 2020. "EPAW: Efficient Privacy Preserving Anonymous Mutual Authentication Scheme for Wireless Body Area Networks (WBANs)." *IEEE Access* 8: 48576–86. doi:10.1109/ACCESS.2020.2977968.

Jing, Qi, Jianbin Hu, and Zhong Chen. 2006. "C4W: An Energy Efficient Public Key Cryptosystem for Large-Scale Wireless Sensor Networks." In *2006 IEEE International Conference on Mobile Ad Hoc and Sensor Systems*, Vancouver, BC, Canada: 827–32. doi:10.1109/MOBHOC.2006.278660.

Joha, M. I., M. M. Rahman, M. S. Nazim, and Y. M. Jang. 2024. A Secure IIoT Environment That Integrates AI-Driven Real-Time Short-Term Active and Reactive Load Forecasting with Anomaly Detection: A Real-World Application. *Sensors*, 24(23): 7440.

Karlof, Chris, Naveen Sastry, and David Wagner. 2004. "TinySec: A Link Layer Security Architecture for Wireless Sensor Networks." In *Proceedings of the 2nd ACM Int. Conf. on Embedded Networked Sensor Syst. (SenSys)*, Baltimore MD, USA: 162–75. https://dl.acm.org/doi/proceedings/10.1145/1031495.

Kothmayr, Thomas, Corinna Schmitt, Wen Hu, and Michael Br. 2012. "A DTLS Based End-To-End Security Architecture for the Internet of Things with Two-Way Authentication." In *Local Computer Networks Workshops (LCN Workshops), 2012 IEEE 37th Conference On*, Clearwater, FL, USA, 956–63. https://www.cse.unsw.edu.au/~wenh/kothmayr_senseapp12.pdf.

Lai, Bocheng, Sungha Kim, and Ingrid Verbauwhede. 2002. "Scalable Session Key Construction Protocol for Wireless Sensor Networks." *IEEE Workshop on Large Scale Real-Time and Embedded Systems (LARTES)*. http://www.emsec.ee.ucla.edu/pdf/2002lartes.pdf.

Li, Xiong, Maged Hamada Ibrahim, Saru Kumari, Arun Kumar Sangaiah, Vidushi Gupta, and Kim-Kwang Raymond Choo. 2017. "Anonymous Mutual Authentication and Key Agreement Scheme for Wearable Sensors in Wireless Body Area Networks." *Computer Networks* 129: 429–43. doi: 10.1016/j.comnet.2017.03.013.

Liang, Gaoqi, Steven R. Weller, Fengji Luo, Junhua Zhao, and Zhao Yang Dong. 2019. "Distributed Blockchain-Based Data Protection Framework for Modern Power Systems Against Cyber Attacks." *IEEE Transactions on Smart Grid* 10(3): 3162–73. doi:10.1109/TSG.2018.2819663.

Liu, Donggang, Peng Ning, and Rongfang Li. 2005. "Establishing Pairwise Keys in Distributed Sensor Networks." *ACM Transactions on Information and System Security* 8(1): 41–77. doi:10.1145/1053283.1053287.

Lu, Yanrong, Lixiang Li, Haipeng Peng, and Yixian Yang. 2015. "An Enhanced Biometric-Based Authentication Scheme for Telecare Medicine Information Systems Using Elliptic Curve Cryptosystem." *Journal of Medical Systems* 39: 1–8.

Maleh, Y., A. Ezzati, and M. Belaissaoui. 2016a. "An Enhanced DTLS Protocol for Internet of Things Applications." In *Proceedings-2016 International Conference on Wireless Networks and Mobile Communications, WINCOM 2016: Green Communications and Networking*, Rabat, Morocco. doi:10.1109/WINCOM.2016.7777209.

Maleh, Y., A. Ezzati, and M. Belaissaoui. 2016b. "DoS Attacks Analysis and Improvement in DTLS Protocol for Internet of Things." In *Proceedings of the International Conference on Big Data and Advanced Wireless Technologies*, Blagoevgrad, Bulgaria. 54:1–54:7. doi:10.1145/3010089.3010139.

Maleh, Y., A. Ezzati, and M. Belaissaoui. 2018. *Security and Privacy in Smart Sensor Networks*. IGI Global. doi: 10.4018/978-1-5225-5736-4.

Maleh, Y., A. Sahid, A. Ezzati, and M. Belaissaoui (eds.).. 2018. "Key Management Protocols for Smart Sensor Networks." *Security and Privacy in Smart Sensor Networks*. IGI Global: 1–23.

Maleh, Y., Y. Qasmaoui, K. El Gholami, Y. Sadqi, and S. Mounir. 2022. "A Comprehensive Survey on SDN Security: Threats, Mitigations, and Future Directions." *Journal of Reliable Intelligent Environments* 9: 201–239. doi: 10.1007/s40860-022-00171-8.

Ming Huang Shun, J., B. Chan, and L. Dai. 2013. "An Efficient Key Management Scheme for Data-Centric Storage Wireless Sensor Networks." *IERI Procedia* 4: 25–31. doi:10.1016/J.IERI.2013.11.005.

Mucchi, L., S. Jayousi, A. Martinelli, S. Caputo, and P. Marcocci. 2019. "An Overview of Security Threats, Solutions, and Challenges in Wbans for Healthcare." In *2019 13th International Symposium on Medical Information and Communication Technology (ISMICT)*, Oslo, Norway. IEEE, 1–6.

Munivel, E., and G. M. Ajit. 2010. "Efficient Public Key Infrastructure Implementation in Wireless Sensor Networks." In *2010 International Conference on Wireless Communication and Sensor Computing (ICWCSC)*, Chennai, India. 1–6. doi:10.1109/ICWCSC.2010.5415904.

Nguyen, K. T., M. Laurent, and N. Oualha. 2015. "Survey on Secure Communication Protocols for the Internet of Things." *Ad Hoc Networks* 32: 17–31. doi: 10.1016/j. adhoc.2015.01.006.

Olufemi, O. O., and D. Adedamola. 2020. "SAMA: A Secure and Anonymous Mutual Authentication with Conditional Identity-Tracking Scheme for a Unified Car Sharing System." *International Journal of Autonomous and Adaptive Communications Systems* 13(1): 84–101.

Perrig, A., R. Szewczyk, J. D. Tygar, V. Wen, and D. E. Culler. 2002. "SPINS: Security Protocols for Sensor Networks." *Wireless Networks* 8(5): 521–34. doi: 10.1023/A:1016598314198.

Pirmoradian, F., M. Safkhani, and S. M. Dakhilalian. 2023. "ECCPWS: An ECC-Based Protocol for WBAN Systems." *Computer Networks* 224: 109598. doi: 10.1016/j. comnet.2023.109598.

Rabin, M. O. 1978. "Digitalized Signatures and Public-Key Functions as Intractable as Factorisation." In Richard J. Lipton, David P. Dobkin, Anita K. Jones (eds.), *Foundations of Secure Computations*, 155–68. Massachusetts Institute of Technology. doi: 10.1080/09720529.2013.858478.

Raza, S., H. Shafagh, K. Hewage, H. Rene, and T. Voigt. 2013. "Lithe: Lightweight Secure CoAP for the Internet of Things." *IEEE Sensors Journal* 13(10): 3711–20. doi: 10.1109/ JSEN.2013.2277656.

Raza, S., S. Duquennoy, T. Chung, D. Yazar, T. Voigt, and U. Roedig. 2011. "Securing Communication in 6LoWPAN with Compressed IPsec. In Distributed Computing in Sensor Systems And." In *IEEE Workshops (DCOSS)*, Barcelona, Spain. 1–8.

Rehman, Z. U., S. Altaf, and S. Iqbal. 2020. "An Efficient Lightweight Key Agreement and Authentication Scheme for WBAN." *IEEE Access* 8: 175385–97. doi:10.1109/ ACCESS.2020.3026630.

Seys, S., and B. Preneel. 2002. "Key Establishment and Authentication Suite to Counter DoS Attacks in Distributed Sensor Networks." *Unpublished manuscript, COSIC*.

Shamir, A. 1984. "ID-BasedCryptoSystem.Pdf." In *Crypto'84*, Santa Barbara, California, USA, 47–54.

Shao, M., S. Zhu, W. Zhang, G. Cao, and Y. Yang. 2009. "PDCS: Security and Privacy Support for Data-Centric Sensor Networks." *IEEE Transactions on Mobile Computing* 8(8): 1023–38. doi:10.1109/TMC.2008.168.

Sikder, A. K., G. Petracca, H. Aksu, T. Jaeger, and A. S. Uluagac. 2021. "A Survey on Sensor-Based Threats and Attacks to Smart Devices and Applications." *IEEE Communications Surveys & Tutorials* 23(2): 1125–59.

Sowjanya, K., M. Dasgupta, and S. Ray. 2020. "An Elliptic Curve Cryptography Based Enhanced Anonymous Authentication Protocol for Wearable Health Monitoring Systems." *International Journal of Information Security* 19(1): 129–46. doi: 10.1007/ s10207-019-00464-9.

Steinwandt, R., and A. Suárez. 2011. "Identity-Based Non-Interactive Key Distribution with Forward Security." *Designs, Codes and Cryptography* 64: 195–96. doi: 10.1007/ s10623-011-9486-0.

Turner, S., and T. Polk. 2011. "Transport Layer Security." *IETF, RFC 6176*.

Watro, R., D. Kong, S.-F. Cuti, C. Gardiner, C. Lynn, and P. Kruus. 2004. "TinyPK: Securing Sensor Networks with Public Key Technology." *2nd Workshop on Security of Ad Hoc and Sensor Networks SASN'04* (October), New York. 59–64. doi: 10.1145/1029102.1029113.

Xu, G., Q. Wu, M. Daneshmand, Y. Liu, and M. Wang. 2015. "A Data Privacy Protective Mechanism for WBAN." *Wireless Communications and Mobile Computing* 16: 1746– 58. doi: 10.1002/wcm.2649.

Xu, Z., C. Xu, W. Liang, J. Xu, and H. Chen. 2019. "A Lightweight Mutual Authentication and Key Agreement Scheme for Medical Internet of Things." *IEEE Access* 7: 53922–31. doi: 10.1109/ACCESS.2019.2912870.

Yang, L., C. Ding, and M. Wu. 2013. "Establishing Authenticated Pairwise Key Using Pairing-Based Cryptography for Sensor Networks." In *2013 8th International ICST Conference on Communications and Networking in China, CHINACOM 2013*-Proceedings, Guilin, China. 517–22. doi:10.1109/ChinaCom.2013.6694650.

Yassine, M., and A. Ezzati. 2015. "Performance Analysis of Routing Protocols for Wireless Sensor Networks." In *Colloquium in Information Science and Technology, CIST*, Tetouan, Morocco. doi:10.1109/CIST.2014.7016657.

Yassine, M., and A. Ezzati. 2016. "LEAP Enhanced: A Lightweight Symmetric Cryptography Scheme for Identifying Compromised Nodes in WSN." *International Journal of Mobile Computing and Multimedia Communications* 7(3): 42–66. doi:10.4018/IJMCMC.2016070104.

Zhu, S., S. Setia, and S. Jajodia. 2003. "LEAP: Efficient Security Mechanisms for Large-Scale Distributed Sensor Networks." In *CCS'03: Proceedings of the 10th ACM Conference on Computer and Communications Security*, Washington, DC. 62–72. doi:10.1145/948109.948120.

Zhu, S., S. Setia, and S. Jajodia. 2006. "Leap+." *ACM Transactions on Sensor Networks* 2(4): 500–528. doi:10.1145/1218556.1218559.

Zigbee, A. 2006. "Zigbee Specification." *Zigbee document 053474r13*.

9 Integrating TinyML with Blockchain for Secure IoT Applications

Doaa Omar and Qasem Abu Al-Haija

9.1 INTRODUCTION

IoT is one of the most prominent technological advances in the current technological era, and it has evolved quickly with varied applications, uses, and advantages. It is projected that over 25 billion will be in use by 2030, according to Statista's exports report [1]. This rapid growth, however, is accompanied by serious challenges, among which security and privacy stand out as the most critical. This is particularly due to these devices' limited resources, weak power, processing capabilities, and memory capacity [2]. Following the development of artificial intelligence and its widespread application in numerous domains, the idea was to merge it with these smart networks to improve their intelligence and effectively tackle security challenges. However, the operational constraints previously mentioned stood in the way of this integration. Around the mid-2010s, TinyML (Tiny Machine Learning) emerged and has gained notable popularity in research and applied fields since 2019–2020 [3]. TinyML enables the deployment of compact machine learning models that run on resource-constrained devices, such as low-power microcontrollers.

Figure 9.1 highlights the key factors that emphasize the need for TinyML technologies in IoT devices. TinyML enables devices to learn and make intelligent decisions in real time, all on their own, without needing to connect to the cloud [4]. TinyML also helps protect privacy since it handles data directly on the device. This means less data is sent wirelessly, a major point of vulnerability in many IoT applications. This method helps save energy, especially in systems where battery life matters. That's because sending data wirelessly usually uses more power than just processing it locally. As a result, companies can reduce the costs related to frequent charging or network usage. TinyML also supports scalability and accommodating small devices with limited resources.

Conversely, blockchain was developed to respond to the accelerated growth of dispersed and decentralized applications and their accompanying security concerns as a sound solution for scalability and data security [5]. Blockchain is commonly viewed as a foundational technology in the current age. It provides a decentralized and secure data storage, exchange, and processing framework.

Blockchain follows the model of a decentralized ledger that records and verifies transactions sequentially using powerful encryption methods. This dispenses with the

DOI: 10.1201/9781003544449-9

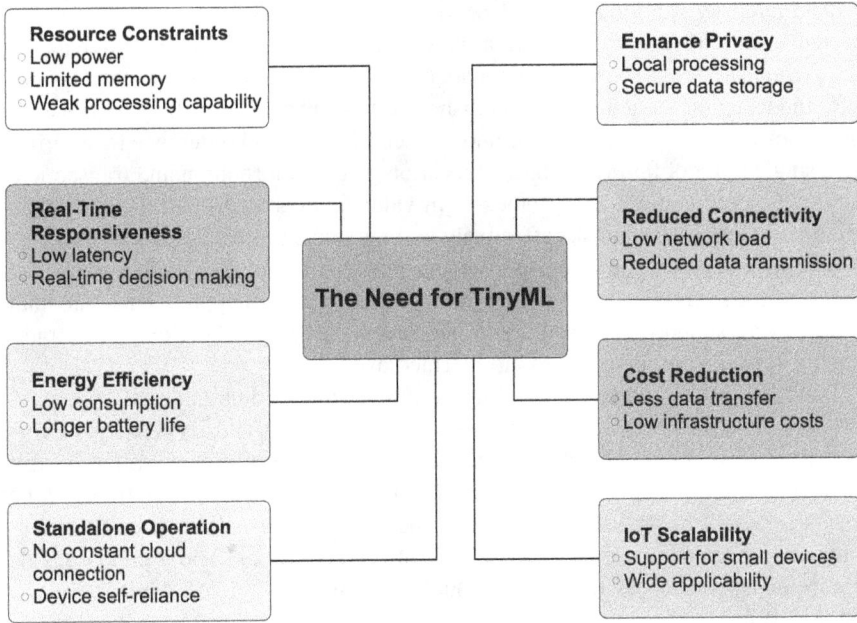

FIGURE 9.1   Key drivers motivating the adoption of TinyML in resource-constrained IoT devices.

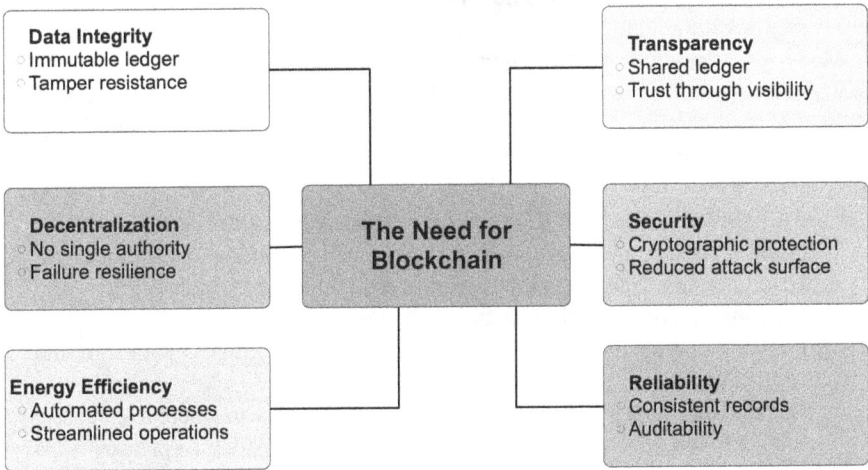

FIGURE 9.2   Core motivations for adopting blockchain technology.

requirement for a middleman, like a bank, in the conventional financial system. Take IBM Blockchain World Wire, for instance, a well-known application in this space, which is a solution by IBM that enables financial institutions to send money directly to one another without relying on intermediaries [6]. As shown in Figure 9.2, blockchain technology addresses several core requirements essential to decentralized systems.

Data integrity is a top issue in decentralized systems. Blockchain addresses this through a shared ledger that maintains records as accurately and tamper-free as possible, rendering the system more dependable [7]. Transparency is also the top attribute, and using the shared ledger establishes participant confidence. Decentralization shares power with many people, promoting security and fault tolerance by eliminating a single point of failure. Robust encryption guards information and transactions, imposing safety and rendering blockchain valuable outside decentralized systems. Smart contracts perform tasks effectively without human interaction, with uses such as peer-to-peer energy trading, permitting direct consumption trade among consumers without middlemen. A Study [8] highlighted several applications that smart contracts can support, such as peer-to-peer energy trading, where energy is traded directly between consumers without intermediaries.

Traditional blockchain technology requires high computational resources; however, lightweight blockchain platforms such as IOTA and Hyperledger Fabric have been developed. The IOTA platform relies on an approach called Tangle, where each new transaction verifies one or more previous transactions, rather than grouping transactions into blocks and adding them all at once to enter a consensus mechanism, as is the case in traditional systems. The Tangle does not require heavy resources like mining, making it suitable for IoT devices running TinyML [9]. Based on the intelligence and capabilities of TinyML technology for resource-constrained devices and blockchain's ability to address security challenges and manage data in these applications, the motivation to integrate TinyML with blockchain has emerged. This integration provides an important chance to overcome challenges in sensitive fields like healthcare, industrial monitoring, and smart city development.

Many recent studies have highlighted the importance of this combination and its contribution to addressing the security challenges hindering the widespread adoption of the Internet of Things. The following points highlight the motivations for integrating TinyML with blockchain:

- **Overcoming Technical Limitations:** Most IoT devices lack computing power, memory, or energy. TinyML offers a smart solution by running lightweight machine learning models that fit well within these constraints.
- **Boosting Security and Trust:** Blockchain works like a reliable digital notebook that keeps data safe and ensures nobody can mess with it in distributed systems.
- **Privacy and Data Protection:** By handling data close to where it's created, this approach cuts down the risk of leaks and helps stick to privacy rules more easily.
- **Support for Scalability and Autonomy:** This approach lets systems grow and adapt more easily without depending fully on a central hub. It also enables them to work more independently and flexibly, especially in different locations.

This chapter aims to comprehensively study integrating TinyML and blockchain to develop secure and efficient Internet of Things applications.

9.2 BACKGROUND AND RELATED WORK

9.2.1 Overview of TinyML

Tiny Machine Learning (TinyML) is a recent branch of machine learning that focuses on running lightweight models directly on resource-constrained IoT devices. It enables intelligent processing on the edge without relying heavily on cloud resources. This technology aims to bring artificial intelligence to the network edge with as little wireless transmission as possible. This approach processes data locally, creating a system that responds in real-time, uses reasonable energy, and reduces privacy risks [10]. The technical characteristics of TinyML have played a key role in enabling the applications mentioned earlier. These characteristics are illustrated in Figure 9.3.

TinyML's key feature is its small model size, which is ideal for devices with limited resources like microcontrollers. Models are usually under 1 MB, which is rare, so advanced techniques help shrink them:

- **Compression:** Makes models smaller by reducing storage needs while stabilizing accuracy. It cuts redundant data and uses efficient representations.
- **Quantization:** Lowers the bit-width for weights and numbers (e.g., 32-bit to 8-bit) to save memory with minimal performance loss.
- **Weight Pruning:** Removes less important weights that have little impact on predictions to slim down the model.

Some machine learning models, like federated learning, work well with TinyML. Federated learning trains simple datasets efficiently while keeping good accuracy. What sets TinyML apart is its excellent power efficiency, which is crucial for devices with limited resources. Studies show that sending data over networks uses much

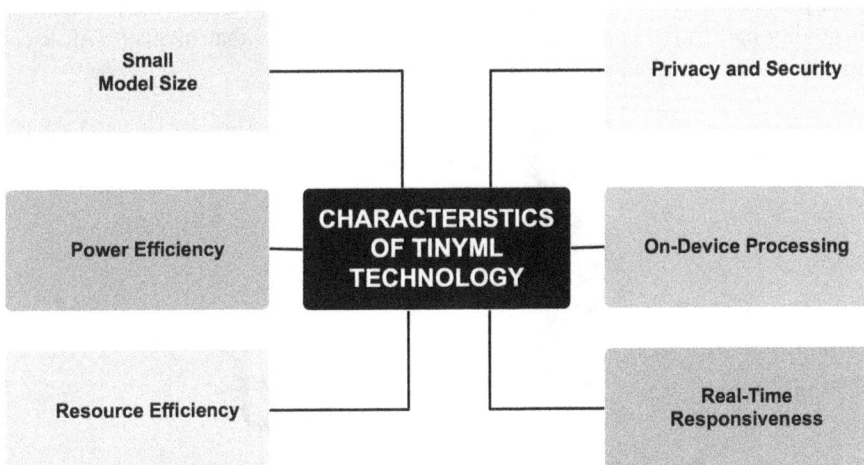

FIGURE 9.3 Core technical characteristics of TinyML technology enabling efficient on-device processing.

more energy than processing it locally on edge devices. That's why TinyML focuses on running models directly on these devices.

With AI demand rising on resource-limited IoT devices, efficient resource use is vital, including power, memory, storage, and processing [11]. TinyML supports this by optimizing models for such constraints. Real-time response is essential in IoT applications like smart home security and industrial monitoring [3]. TinyML is highly flexible, making it easy to apply in different environments with minimal adjustments. This reduces the need to create separate models for each case. A 2025 study published on medRxiv demonstrated that an ECG model built with TinyML and trained on one dataset was still accurate when used on new data, even with varying conditions and patient differences [12]. TinyML applications span a wide and diverse range of fields; we will cover some of them.

TinyML is widely used in the Industrial Internet of Things (IIoT) and intelligent transportation and motion monitoring systems. One study proposed a solution in this field based on edge devices as a practical example. The goal was to improve road safety by monitoring the driver's condition in real time to detect signs of drowsiness or fatigue. Researchers used a convolutional neural network (CNN) quantization model to compress the model size to approximately 0.05 megabytes. The lightweight nature of the model enabled fast real-time analysis, with response times in the range of a few hundred milliseconds [13].

Healthcare is a key area for TinyML, enabling real-time processing of vital signs via wearables. In one study, a microcontroller (64 MHz CPU, 256 KB RAM) ran CNN models with sensors collecting multidimensional data. The goal was to detect cardiac arrhythmias in real time (Figure 9.4). Quantization and pruning were used to fit device constraints, reducing weights from 32-bit to 8-bit and removing low-impact nodes to cut computation. This shows TinyML's promise in continuous monitoring and early detection of conditions like arrhythmias through real-time alerts [14]. TinyML also includes a wide range of modern applications that rely on the intelligent processing distribution. This includes smart cities that use sensors to monitor air quality and traffic congestion and smart home devices that interact with users through voice or motion, among other applications.

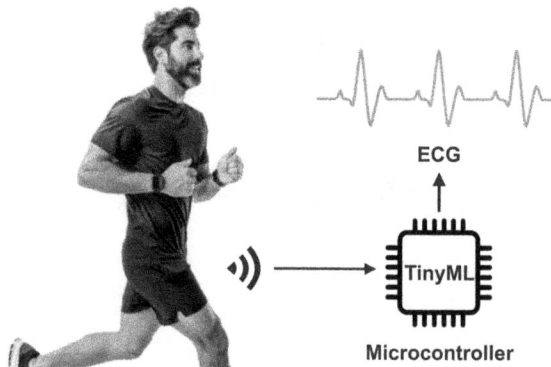

FIGURE 9.4 Application of TinyML in enhancing healthcare system efficiency and real-time data processing.

Furthermore, it extends to agriculture, where artificial intelligence is integrated to monitor soil conditions and crop health. In this context, edge computing was utilized for plant disease detection using a lightweight model, specifically MobileNetV3-small. Using quantization techniques, the model size was reduced from 1.5 million to 0.93 million, making it suitable for deployment on edge devices within real-world agricultural environments. The results highlight the effectiveness of TinyML in running accurate models for real-time plant disease detection, even in rural areas with limited connectivity, achieving an accuracy of 99.5% [15].

Implementing a TinyML model relies on a set of integrated tools and frameworks. These are used in multiple stages, including training, optimization, conversion, and deployment on edge devices.

Table 9.1 illustrates the implementation of a TinyML model and the key tools required for each stage.

Various tools support TinyML model development from training to edge deployment. **TensorFlow** is used to train models on computers. **Edge Impulse** is a cloud platform with a user-friendly interface for data collection, training, and performance

TABLE 9.1

Overview of the Implementation Stages of a TinyML Model and the Key Tools Required at Each Stage

Stage	Description	Tool or Framework	Purpose
1. Model training	The model is initially trained on a standard computer using collected data.	• TensorFlow (full version) • Edge Impulse	To build and train a machine learning model, Edge Impulse simplifies this process via a graphical interface.
2. Model optimization	Once trained, the model is usually too large, so it must be reduced to run efficiently on a small device.	• TensorFlow Lite • MicroTVM • Quantization tools	To reduce model size and computational cost, enabling deployment on constrained devices.
3. Model conversion	The optimized model is converted into a format compatible with the target hardware.	• TensorFlow Lite for Microcontrollers • Edge Impulse Exporter	To generate deployable formats such as .tflite or C arrays compatible with microcontrollers.
4. Deployment on device	The converted model is embedded into a microcontroller program and connected to real-time sensor inputs.	• Arduino TinyML Libraries • CMSIS-NN	To integrate the model with sensor input/output and ensure efficient execution on low-power processors.
5. On-device inference	The device runs the model in real time to make predictions from live data, fully offline.	Same tools as above	To perform live inference locally, without relying on cloud services or external computing power.

analysis. **TensorFlow Lite** converts models to smaller versions for low-resource devices. **Quantization** tools reduce weight precision (32-bit to 8-bit) to lower size and power use. **MicroTVM** optimizes model execution on low-power hardware. **TFLite for Microcontrollers (TFLM)** runs models directly on devices like Arduino without an OS. **Edge Impulse Exporter** outputs models in formats like .tflite or C arrays. **Arduino TinyML** Libraries help run models on Arduino boards. **CMSIS-NN** offers fast neural network functions for ARM Cortex-M processors.

In the following section, we examine the fundamental principles of blockchain technology and discuss its significance and applications within the domain of the Internet of Things.

9.2.2 BLOCKCHAIN BASICS FOR IoT

The Internet of Things involves millions of devices exchanging data that must be stored securely. Blockchain supports this by creating a trusted, decentralized system without central control. Acting as a digital ledger, it records transactions securely and sequentially. Data is distributed across multiple devices, preventing central dominance. This design ensures data integrity and protection from tampering, which is crucial in IoT environments. To grasp the technical pillars that support blockchain applications within an Internet of Things environment, you may refer to Figure 9.5, which outlines the four core concepts that serve as a foundation for understanding this technology.

After reviewing the core concepts of blockchain technology, as presented in Figure 9.7, it becomes important to complement this visual overview with a deeper analytical reading.

Distributed ledgers form blockchain's core, tracing real-time data from IoT edge devices like sensors. Data is shared across many nodes, boosting transparency and preventing tampering. This is used in smart supply chains to track products from

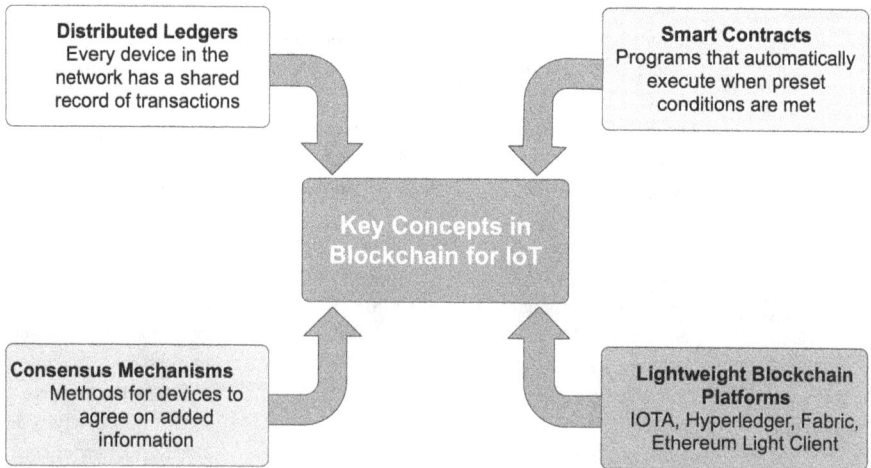

Distributed Ledgers
Every device in the network has a shared record of transactions

Smart Contracts
Programs that automatically execute when preset conditions are met

Key Concepts in Blockchain for IoT

Consensus Mechanisms
Methods for devices to agree on added information

Lightweight Blockchain Platforms
IOTA, Hyperledger, Fabric, Ethereum Light Client

FIGURE 9.5 Key concept of blockchain.

production to consumption. Smart contracts are programs that execute automatically when conditions are met, such as verifying health insurance before medical procedures, speeding up processing [16]. IoT's limited power challenges consensus methods like Proof of Work (PoW), so alternatives like Delegated Proof of Stake (DPoS) use trusted validators to verify transactions efficiently, saving energy [17]. Platforms like IOTA, using DAG instead of blockchain, handle small, frequent sensor transactions with no fees and fast response, as shown in smart transport trials [18]. Hyperledger Fabric offers customizable private blockchains with access control, which are used in medical and industrial trials by IBM and its partners [19].

Despite the tremendous potential these mechanisms and frameworks offer, the question that lingers is: Are these ideas universally applicable in Internet of Things applications, or is the outcome determined by the particular context and the attendant technological issues involved? Theoretically, the answer is yes. Practically, the feasibility of deployment depends on the infrastructure, the nature of the device used, and the sensitivity of the data to which the system is being applied. Remote applications, for example, in agricultural systems, or highly secure applications such as ICU systems in hospitals, can face real problems using conventional blockchain systems due to limitations in available technology assets and the urgency of the need to respond to events in real time. For this reason, understanding the theoretical concepts alone is not enough; what matters more is how well the proposed solution aligns with the target system's specific nature and actual needs. In the subsequent sections, these ideas will be presented in more detail, emphasizing the envisioned architectural models, the mechanisms of integrating TinyML with blockchain, and the technological issues of their implementation in real applications, particularly in edge devices with constrained resources.

9.2.3 SECURITY CHALLENGES IN IoT

With the increased spread of the Internet of Things across different sectors of life, starting from homes to manufacturing facilities and transportation systems, there is also a growing need to secure these systems against a broad spectrum of threats, as the threats have the potential to compromise user anonymity and data integrity. The inherently open and constantly connected nature of the Internet of Things makes it an attractive target for cyberattacks, especially when devices are small, resource-limited, and not regularly updated [20].

Among the most common threats are **spoof attacks**, in which the adversary impersonates a genuine device granted access to data communication channels in a manner not authorized by the original user or users. If we take the example of home smart networks, for instance, a hacker might be able to manipulate a thermostat or security camera once the initial spoofing has passed [21]. **Data tampering** poses an equally severe threat, in which an attacker manipulates the data flowing from or into devices. Such an attack tends to be sneaky and hard to detect, as it can be used in industrial or healthcare settings, where a small sensor value adjustment can cause the wrong system decision [22].

Protecting access to devices and data is a major challenge in many IoT environments, especially when these systems lack strong security measures or rely on weak, easily guessed passwords. Under such circumstances, the system will be exposed to **unauthorized access attacks**, either caused by misconfigurations or weak communication channel encryption, thereby providing the attackers with an easy access point into the system and modification of the information [22].

Although the variety of security threats that IoT environments may face is significant, we have chosen to focus on the previously mentioned threats due to the severity of these attacks and their direct impact on data and system integrity.

Even with legacy security solutions such as encryption, two-factor authentication, or firewalls, their applicability in the Internet of Things is limited. These devices often do not have the resources to perform sophisticated protocols because their memory and power resources limit them. For instance, a low-power controller in a remote sprinkler system cannot do advanced encryption like a personal computer. Furthermore, centralized servers for managing identity or integrity verification introduce points of failure and weaken the system's resilience in facing focused attacks [23].

From these points, the need grew stronger for solutions that bridge the gaps and tackle the challenges in this field. Such solutions must consider the limitations of the devices and avoid placing excessive strain on the system. Among the promising recent trends is blockchain technology, with its features for decentralized verification and early threat detection using localized TinyML models. These two areas will be explored in greater detail later in this chapter.

9.2.4 RELATED WORK

To be more informed about the most prominent contributions in this field, it is essential to review the research works that targeted the application of TinyML technologies and blockchain and their integration. Previous studies covered a wide range of applications that contributed to addressing many challenges and offered innovative solutions using these technologies. Table 9.2 provides a summary of the most prominent of these studies.

The table only partly shows the variety of applications and research in these fields. For example, study [24] highlights TinyML's effectiveness in healthcare wearables like smartwatches, using low power (about 0.17 watts) while maintaining high energy efficiency. However, limited device resources may restrict full data security. Blockchain ensures secure data exchange among devices. Studies [26] (2024) highlight the key role of blockchain in smart city traffic management, where smart contracts automate traffic rules to enhance decision-making and traffic flow without compromising security. Yet, blockchain faces challenges with IoT devices' energy and processing limits, affecting speed. Integrating TinyML and blockchain offers smarter, more secure IoT systems. TinyML enables local machine learning on limited devices, while blockchain provides data integrity and decentralization. This integration allows faster, more reliable cyberattack responses with less cloud dependence, enhancing system autonomy and security, as shown in studies [29–31].

TABLE 9.2
Summary of Previous Works in the Field, Highlighting Key Findings and Contributions

References	Domain	Methodology	Key Points	Results	Notes
Zaidi et al. [24], IEEE Access, 2022	TinyML	TinyML-as-a-Service architecture	Framework for running ML on resource-constrained devices	Low power consumption, gesture recognition use case	Challenges in transfer and federated learning
Suwannaphong et al. [25], Sci. Rep., 2025	TinyML	Compressed models (quantization and distillation)	Model size reduction while maintaining localization accuracy	Effective in low-memory environments, a healthcare application	Transformer and Mamba models
Huang et al. [12], medRxiv, 2025	TinyML	μ-Training and μ-Fine-Tuning for small models	Secure and efficient local training on edge devices for ECG data analysis	High computational efficiency, privacy protection, and superior performance	Biomedical application and ECG signal analysis
Alajlan et al. [13], Sensors, 2023	TinyML	TinyML-based driver drowsiness detection using deep learning	Lightweight DL models using TinyML for driver drowsiness detection	Small model size and high detection accuracy	Performance improvement using quantization techniques
Bandhu et al. [7], Multimedia Tools and Applications, 2023	Blockchain	Ethereum blockchain-based drug supply chain	Using smart contracts to track drugs across the healthcare supply chain	Transparency, drug data tracking, and low gas cost	Compared with previous solutions, no off-chain data storage
Chen et al. [26], J Grid Computing, 2024	Blockchain	Smart Traffic Management System (STMS) with Blockchain, IoT, and Edge Computing	Integration of Blockchain, IoT, Edge Computing, and TD3 for smart traffic management	Improved traffic flow, reduced congestion, increased efficiency with security and transparency	Innovative smart city traffic management using smart contracts

(Continued)

TABLE 9.2 *(Continued)*
Summary of Previous Works in the Field, Highlighting Key Findings and Contributions

References	Domain	Methodology	Key Points	Results	Notes
Aljumah [27], Scientific Reports, 2025	Blockchain	Distributed security framework for IoT networks using SDN, blockchain, and edge computing	Integration of SDN, Blockchain, and edge computing for attack detection and prevention in IoT networks	Detection accuracy 98.7%, false positive rate 1.2%, fast response (101.1 ms)	Enhanced security in complex IoT environments using the Ethereum network
Shalan et al. [28]	TinyML + Blockchain	Federated learning framework with blockchain for intrusion defense in smart homes	Combining federated learning with knowledge distillation and blockchain-based RBAC access control	Enhanced security, data privacy, resource efficiency, and scalable smart networks	High-accuracy practical evaluation on N-BaIoT, advanced access control mechanisms, and support for dynamic environments
Tsoukas et al. [29]	TinyML + Blockchain	Food supply chain security system using blockchain and TinyML	End-to-end integrated monitoring system, anomaly detection using TinyML, enhanced traceability, and transparency	Improved food supply chain security, fraud reduction, and increased trust	Technology integration to reduce attacks and improve traceability and transparency
Adhikary et al. [30]	TinyML + Blockchain	TinyWolf: On-device TinyML training using enhanced gray wolf optimizer	Training TinyML models on resource-constrained devices using a nature-inspired optimization algorithm	Efficient training with low memory and processing consumption	Application on microcontrollers (256 KB RAM) with improved memory usage

9.3   ARCHITECTURAL OVERVIEW OF TINYML-BLOCKCHAIN INTEGRATION (1,200–1,500 WORDS)

9.3.1   System Design Goals

In systems that integrate TinyML with blockchain for the Internet of Things, the primary design goals focus on balancing several key pillars. These include efficiency, security, scalability, and decentralization.

Efficiency refers to the system's ability to effectively meet its core requirements while minimizing energy and resource consumption to enable IoT devices to run TinyML models locally without heavy downloads or excessive resource consumption. Security involves ensuring the integrity and accuracy of data both during transmission and storage, achieved through blockchain technologies that prevent tampering. The integration also helps detect attacks early by running machine learning models directly on devices using TinyML technology.

System flexibility and its ability to adapt to changes in the number of devices and network expansion are also fundamental pillars. Integration ensures this capability. This is because TinyML enables local operation on small and simple devices. Blockchain can easily scale the network by adding new nodes without the complexities of centralized control. Achieving decentralization helps reduce single points of failure, as tasks are distributed across multiple nodes, the principle underlying blockchain technology. TinyML also supports this by enabling small devices to perform intelligent tasks locally.

9.3.2   Overall Architecture

Let us look at the system's overall architecture, which integrates TinyML and blockchain. Refer to Figure 9.6. The system comprises multiple connected layers designed to perform a specific role. They seamlessly integrate to fulfill the application's required functions intelligently.

In the sensing layer, data is collected from the environment. As shown in the diagram, the data exiting this layer is represented in form of spreads, which suggests that it is raw and requires further processing.

The edge processing layer, consisting of small, resource-constrained devices, resembles IoT devices capable of running lightweight TinyML models. These edge devices analyze the raw data received from the sensing layer according to the purpose of the application, using the deployed machine learning models. TinyML models perform inference on the collected raw data, detecting anomalous patterns, classifying events, or generating immediate security alerts. After processing, the important results are sent through the communication layer to higher levels, where the data is managed on a broader scale and represented with an orange color, symbolizing alerts or significant events detected following the analysis.

The communication layer is one of the vital components of the overall architecture of these systems to ensure smooth and efficient data flow. It applies multiple protocols and technologies designed to meet the distinct needs of IoT environments. For example, data transmitted from the sensing layer to the edge layer typically

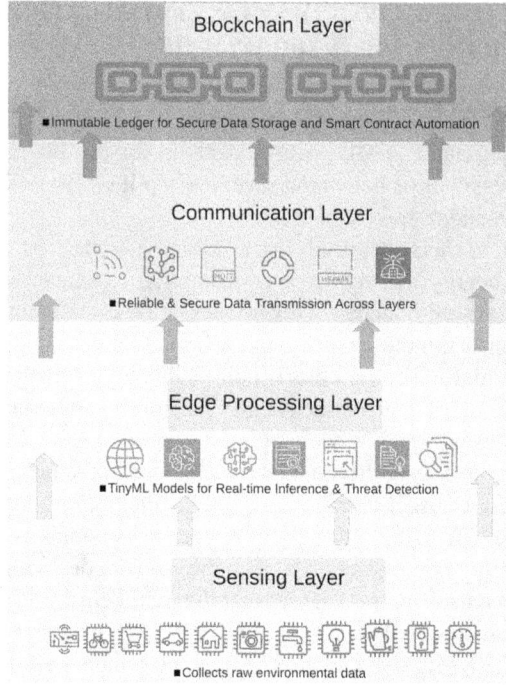

FIGURE 9.6 Integration of TinyML and blockchain, highlighting key component interactions.

relies on protocols such as Wi-Fi, Zigbee, and LoRa, which are well suited for efficient short-range communication. Data flow from the edge to the blockchain layer typically uses 4G and 5G networks. In the blockchain layer, final verification and recording processes occur, where the processed and trusted data is recorded in an immutable distributed ledger. When inference results in an event or alert that requires action, these outcomes are sent to the blockchain layer, where validation takes place to ensure the accuracy and integrity of the data. Through smart contracts that operate automatically on the blockchain network, the processed data is verified for accuracy according to predefined rules, and automatic actions are executed accordingly.

Note: TinyML can be deployed directly on IoT devices themselves, rather than only on edge devices, provided that their specifications and capabilities support running lightweight machine learning models.

9.3.3 COMPONENTS AND INTERACTIONS

In this section, we will explore the system's main components that integrate TinyML and blockchain and how these components interact with one another. Figure 9.7 illustrates how the system components integrate and work together.

The sensing layer consists of IoT devices that collect information from the environment. These are typically small devices such as security cameras, temperature

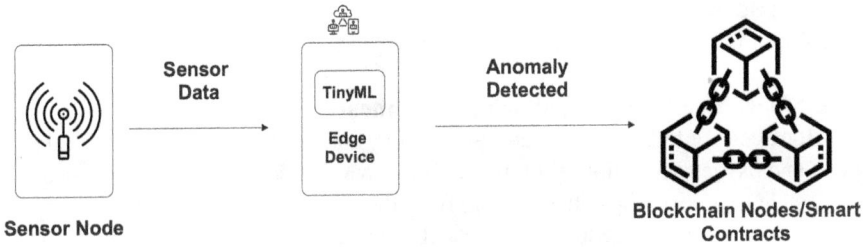

FIGURE 9.7 Main system components and data flow.

and humidity sensors, and smartwatches. They collect raw data from the surrounding environment and can be equipped with lightweight TinyML models that run locally on the hardware, as highlighted by numerous studies [31].

The edge processing layer is also a resource-constrained device capable of running lightweight TinyML models. These devices represent the analytical core of the system, where they are relied upon to run pre-trained TinyML models, enabling the system to meet specific application requirements. Edge devices receive data from multiple sensor nodes and perform inference operations locally. When a significant or anomalous event is detected, the device generates a report containing the analysis result and transmits it to the blockchain network. These devices may also include additional logic to filter data, reduce redundancy, and manage logging priorities. After edge devices complete the data analysis, the inference results are passed to the blockchain network for documentation and verification, where the significant events are recorded in a distributed ledger. Smart contracts on the blockchain enable the system to handle events, process data, and make decisions accordingly without manual intervention.

To maintain efficiency and compatibility with the constrained hardware capabilities of edge devices, lightweight blockchain solutions such as Ethereum Light Clients or Directed Acyclic Graph (DAG) networks like IOTA are often preferred [32]. To better understand the concept, TinyML and blockchain should be used to monitor industrial production lines and identify minor faults that might otherwise go unnoticed. In this context, precise sensors are deployed on machines to capture acoustic and vibration data, aiming to detect potential faults in the equipment. The collected data is sent to nearby edge devices equipped with trained TinyML models specialized in fault detection and capable of identifying subtle patterns or micro-anomalies that are not visible to the naked eye. For example, the model might detect an unusual vibration frequency in one of the rotational axes, indicating the early stages of internal wear in a delicate mechanical component. Upon detection, a real-time alert is generated within the industrial environment, and the outcome is securely logged onto the enterprise's private blockchain ledger. Such records can later be accessed for quality assurance reviews or investigative purposes in case of defects or anomalies in the final product.

9.3.4 DEPLOYMENT SCENARIOS

The integration of TinyML and blockchain represents a significant leap in the infrastructure of Internet of Things (IoT) applications. A balance between real-time data analysis and tamper-proof security is achieved. This model can be applied across a wide range of sectors, each facing its unique challenges and benefiting from different capabilities offered by the integrated system.

1. **Smart Healthcare Monitoring (Healthcare IoT):** Wearable healthcare devices are tools that collect real-time physiological data such as heart rate and oxygen saturation. With the use of TinyML models, it has become possible to process the data collected from these devices locally without a long transition, which reduces the risk of data being stolen or tampered with. Moreover, using blockchain networks to record sensitive medical events ensures integrity and tamper resistance and protects high-value informational assets, especially since the average selling price of medical data on the black market exceeds that of regular personal data by a factor of 50 [33]. When an out-of-range medical event is detected (such as tachycardia or sleep apnea), an encrypted alert is generated and sent through a secure channel, with the outcome recorded on a dedicated medical blockchain network. This model ensures patient privacy while providing an immutable record of any detected health emergency (see Figure 9.8).

2. **Industrial Internet of Things (IIoT):** This integration has enhanced the quality of machine and process monitoring in factories and industrial facilities, reducing downtime and improving operational efficiency. Sensor arrays are strategically deployed to detect mechanical movements, temperature variations, or vibration patterns, depending on the specific monitoring requirements. Collected signals are transmitted through energy-efficient communication interfaces to integrated edge devices equipped with TinyML models trained to recognize the normal operational patterns of the machines, capable of performing real-time inference to detect early signs of malfunction (Early Fault Detection). Upon surpassing predefined critical thresholds, the system triggers the creation of event logs, which are securely stored on the organization's private blockchain for traceability and audit purposes [34] (Figure 9.9).

3. **Smart Cities and Infrastructure Systems:** In urban environments, monitoring traffic flow, air quality, public lighting systems, and critical infrastructure are among the most essential functions of smart cities. Strategically positioned edge devices leverage TinyML models to perform real-time analysis of environmental phenomena, including traffic flow intensity, water leakage detection, and air quality indicators. Operationally

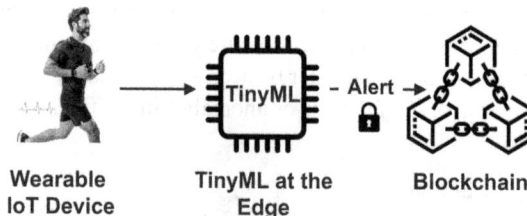

| Wearable | TinyML at the | Blockchain |
| IoT Device | Edge | |

FIGURE 9.8   Remote health monitoring using a wearable device, TinyML at the edge, and data logging via blockchain.

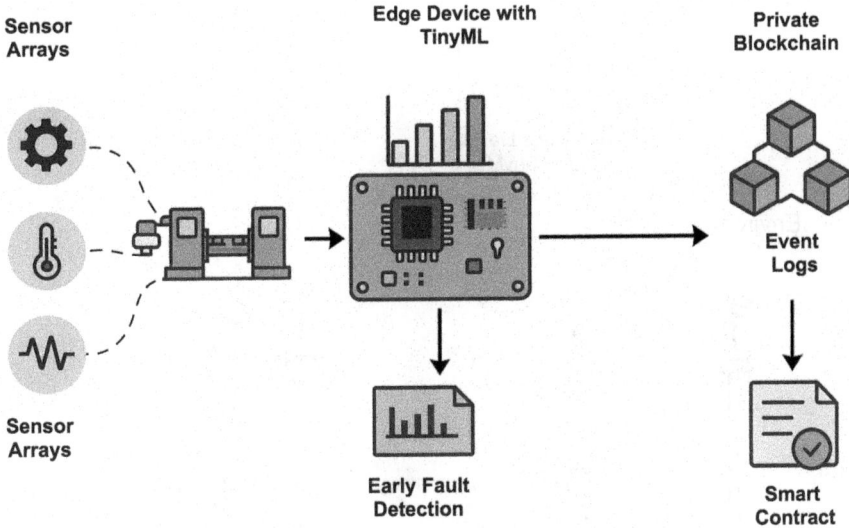

FIGURE 9.9 TinyML and blockchain integration in industrial environments for early fault detection and secure logging.

significant results are recorded on a blockchain network, enabling precise historical analysis while ensuring data integrity. Smart contracts also adjust traffic signals, reroute vehicles, or control adaptive lighting based on environmental conditions. Refer to Figure 9.10.

4. **Precision Agriculture:** In agricultural settings, implementing precision farming, key factors such as soil quality, moisture content, and pest presence are continuously monitored. Drones and ground sensors transmit data to mobile edge units running TinyML models trained to detect water stress, uneven growth, or pest infestations. The decision execution record is logged on an agricultural blockchain to foster transparency between producers and distributors and to ensure compliance with environmental and agricultural standards (Figure 9.11).

5. **Smart Supply Chains:** In supply chains, smart containers are monitored using sensors that track physical variables such as temperature, humidity, and vibrations. TinyML units on edge devices analyze this data locally to detect deviations from ideal conditions, such as a break in the cold chain or excessive product vibration. In the event of a fault, an incident log with geotemporal data is sent to a logistics blockchain network, which enables real-time product tracking and provides transparent proof of liability in cases of shipment quality degradation [29] (see Figure 9.12).

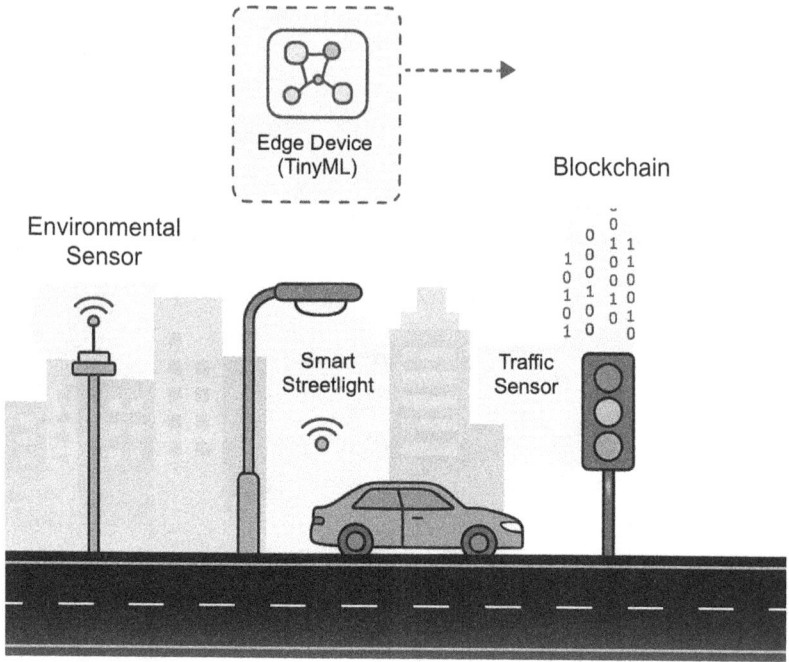

FIGURE 9.10   Smart city infrastructure integrating TinyML edge devices with blockchain for real-time traffic and environmental data analysis.

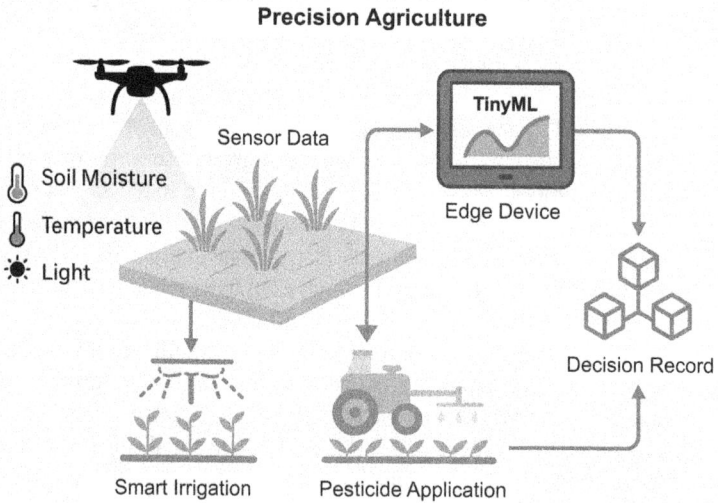

FIGURE 9.11   Precision agriculture using TinyML and blockchain for smart irrigation and pesticide control.

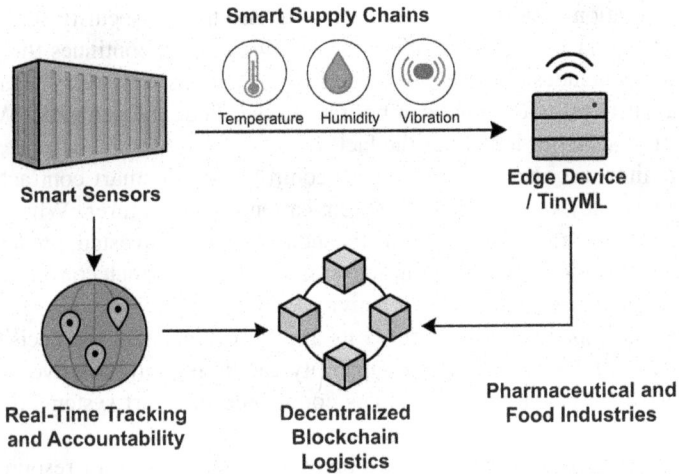

Smart Supply Chains

Temperature   Humidity   Vibration

Smart Sensors

Edge Device / TinyML

Real-Time Tracking and Accountability

Decentralized Blockchain Logistics

Pharmaceutical and Food Industries

FIGURE 9.12   Smart supply chain system using TinyML for anomaly detection and block-chain for secure, real-time tracking.

9.4   SECURITY AND PRIVACY ENHANCEMENTS VIA INTEGRATION

Resource-constrained devices are widely distributed across large, heterogeneous networks in the IoT environment. Security and privacy requirements have become more complex than ever before. Integrating TinyML and blockchain technology combines real-time analytical and intelligent response capabilities with a tamper-resistant infrastructure built on decentralized trust. Here, we examine how the integration contributes to more effective threat detection, robust data protection, and the facilitation of precise forensic analysis while addressing the accompanying technical challenges.

Deploying TinyML models on edge devices enables data analysis and inference to be performed directly at the node level. This enables immediate response upon detecting suspicious behavior or abnormal patterns, such as a sudden change in vibration frequencies of an industrial device or unexpected movement patterns in a smart home environment, as inference is performed within a single millisecond in many cases, enabling the system to trigger security responses, such as locking a door or issuing an alert in real time. By enabling local data processing, the approach limits external transmission and reduces the potential for data interception or manipulation, effectively shrinking the attack surface.

Blockchain ensures immutability and trust by providing a distributed ledger structure that cannot be altered once data is written. This ledger structure chronologically and geographically logs the results of security inferences with precision [35]. When a security event is detected at the device level, its details are recorded on the blockchain as a digitally signed transaction. This creates an immutable and verifiable historical record for future audits and investigations, which proves ideal for ensuring the integrity and protection of security event records.

This integration enables what is known as end-to-end security integration. It begins at the point of data generation on the edge device, continues through local analysis, and concludes with decentralized logging. In this way, every security event becomes a trustworthy unit that can be tracked, verified, and analyzed. Whether at the moment it occurs or long after the fact.

This architecture also supports access control through smart contracts, removing the dependency on centralized authentication infrastructures. When access to specific data or the activation of a sensitive component is requested, predefined conditions are automatically verified through smart contracts, which consider user identity, geographic location, or previous activity log [36].

One notable aspect of this model is its ability to document the decisions made by TinyML AI models [37]. Such capability facilitates retrospective auditing of machine-driven decisions and reinforces confidence in smart systems deployed in critical settings.

Despite all these benefits, applying blockchain technology to resource-limited device settings remains hugely technically challenging because of the requirement of conventional consensus algorithms employed in blockchain, like Proof of Work or Proof of Stake [38]. This necessitates the investigation of lighter alternatives compatible with the limitations of such an environment.

One option is to use light client models, which allow a device to verify transactions by storing only block headers rather than the entire blockchain [39]. Another method involves off-chain ledger designs, where large amounts of data are kept outside the blockchain while only their hash or digital signature is stored within the block [40]. Networks that move away from conventional chain-based structures, like IOTA, built on a directed acyclic graph (DAG), also offer practical solutions that support efficiency and adaptability [41].

The convergence of TinyML and blockchain illustrates that security has evolved beyond being merely an infrastructural consideration; it is now embedded across all system layers, from edge devices to the highest levels of the network. The model proves highly effective in real-world scenarios where rapid security, autonomous actions, and traceable documentation are critical. The next section provides a case study demonstrating the practical implementation of this integration within a smart home setting.

9.5 CASE STUDY: SMART HOME INTRUSION DETECTION SYSTEM

With the widespread use of connected devices in smart homes, the need for effective security systems to detect and analyze abnormal activities in real time is growing. In this context, the study [28] will be adopted as a case study to illustrate how TinyML and blockchain can be utilized to enhance the security of smart homes. We will explore how TinyML can be used to analyze data and detect suspicious activities locally in real time. We will also discuss how blockchain can offer an extra level of protection in smart homes by documenting and recording events in an immutable manner and delve into the technical design details of the system, with a focus on the algorithms employed, and examine how to integrate these technologies into the system cohesively and efficiently.

9.5.1   Scenario: Detecting Abnormal Activities in a Smart Home

This system aims to identify abnormal activities within the home by utilizing motion sensors and intelligent cameras. These sensors collect data on movement within the home; upon detection of abnormal behavior, the data is processed through a TinyML model trained to recognize atypical patterns. The data is analyzed in real time on edge devices using TinyML, allowing real-time detection of potential threats. A real-time alert is transmitted to the user interface when suspicious movement is detected, allowing the owner to respond accordingly.

9.5.2   TinyML Model: Person Detection or Motion Pattern Recognition

This system aims to detect abnormal activities within the home using motion sensors and smart cameras. For example, the model can detect the presence of an individual in the home at an inappropriate time or in a restricted area. The data is collected to detect any changes in movement patterns. To perform this analysis, the model relies on federated learning algorithms, an algorithmic approach used to train models across multiple devices or nodes in a distributed manner. Using federated learning allows devices to continuously improve their models while keeping local data away from servers. Although federated learning is lightweight in terms of its ability to process local data across edge devices without transferring data to servers, in some cases, further enhancement in model size is required due to the limited resources of edge devices. For this purpose, knowledge distillation technology has been used. Knowledge distillation reduces the size of a large model while maintaining its good performance, transforming the large model into a smaller one. Then comes federated learning, where the smaller model, reduced using knowledge distillation, is applied across multiple devices. In this case, each device trains the local model using its local data. The updates are sent to a central server after training the local models on the small devices. Thus, a single, powerful, optimized model is created based on learning from multiple devices without sharing data.

9.5.3   Blockchain Usage: Logging Events, Sharing Alerts, and Authentication

In this system, blockchain is considered a fundamental layer to ensure security and credibility in handling the data. The alert is sent to the decentralized blockchain network once abnormal activity is detected through the TinyML model. This enables permanent, tamper-proof documentation of all alerts, enhancing transparency and assisting in subsequent investigations in case of a breach or security threat. The documentation and data storage processes are carried out in a distributed ledger, where each event, including abnormal activities such as intrusion attempts, is recorded in encrypted blocks. When interacting with the system, smart contracts authenticate and verify the data securely, ensuring that these alerts remain tamper-proof in addition to the automatic execution of actions based on detected patterns, such as notifying the homeowner or sending an alert to the police in the event of a significant threat.

- **Hardware Used:** It is worth noting that the system relies on microcontrollers such as the Arduino Nano 33 BLE Sense, which features an ARM Cortex-M4F processor running at 64 MHz and 256 KB of RAM, along with a fully integrated set of sensors, including motion, temperature, and pressure sensors.
- **Performance Metrics—Accuracy, Energy Consumption, Latency:** The models based on federated learning demonstrated high accuracy in detecting abnormal activities, outperforming traditional machine learning models. Thanks to the use of lightweight and efficient algorithms, power consumption was also significantly optimized, which allowed the devices to operate for extended periods on small batteries without transferring data to central servers. Latency was substantially minimized through local data processing, enabling the system to deliver faster responses than conventional systems.

9.6   IMPLEMENTATION CONSIDERATIONS AND OPTIMIZATION (1,000–1,200 WORDS)

Model Compression Techniques: One of the key features of TinyML is its small model size, which is well suited for resource-constrained devices, such as microcontrollers. The machine learning model is expected to be smaller than 1 megabyte, which is rare. Therefore, advanced techniques are often employed to reduce the model size, including:

- **Compression:** Making the model more compact through compression reduces storage demands but tries to keep accuracy steady. Precision reduction can reduce the number of parameters the model needs to store, resulting in a smaller and more efficient model with minimal impact on performance. In doing so, the process also helps eliminate redundancy in the data and encourages the use of more efficient and concise representation methods.
- **Quantization:** A common method for shrinking model size is to lower the bit-width used to represent weights and numerical values, helping to reduce memory usage without major loss in performance. Aim to reduce the precision used to represent the model's weights and other numerical values. For example, instead of using high-precision decimal numbers (32-bit or 64-bit) to represent the weights, the number of bits is reduced to a smaller value (8-bit or 16-bit). In addition to approximating values (such as weights).
- **Weight Pruning:** Pruning involves removing unimportant weights from the model. The pruned weights usually contribute the least to the model's predictions.

9.6.1   Efficient Blockchain Operations: Off-Chain/On-Chain Balance

Blockchain effectively enhances security and transparency, while TinyML is one of the effective solutions that enables edge devices to execute AI models capable of analyzing data and responding in real time. The integration of these technologies presents several challenges, with the primary one being achieving the optimal balance between on-chain and off-chain processes. One of the biggest challenges in integrating blockchain with TinyML is finding the optimal balance between on-chain and off-chain operations. On-chain operations involve directly recording transactions within the distributed ledger or blockchain. This method provides strong security benefits; however, it is often slow and energy-consuming, which conflicts with the efficiency demands of TinyML. In a blockchain system, smart contracts can record transactions on-chain. However, efficient operations such as preliminary data analysis or model validation can be performed off-chain, which allows faster and more efficient data processing. For instance, when employing a TinyML model to analyze motion data, most of the processing occurs locally on the edge device, and only the segmented results are transmitted to the distributed network via on-chain operations.

9.6.2 Data Handling: Periodic Writing versus Event-Based Writing in Blockchain

In systems that integrate blockchain with TinyML on edge devices, massive amounts of data flow periodically from the edge to the blockchain. Choosing the optimal method for writing data to the blockchain is crucial to ensure system efficiency and reduce costs. There are two approaches for this purpose: periodic writing and event-based writing. Periodic writing refers to recording all events in the blockchain, regardless of their value or significance. In this method, it is essential to continuously update the blockchain to ensure the ledger is updated periodically. This increased transaction costs due to the high frequency of writes occurring within brief intervals. On the other hand, event-driven writing records data on the blockchain exclusively when a defined event takes place, including instances like abnormal motion detection or security threat identification. This method improves system efficiency by logging only critical events, reducing transaction frequency, and processing costs.

9.6.3 Energy and Memory Consumption Optimization Strategies

In the Internet of Things systems, where edge devices rely on limited batteries, energy and memory consumption are critical to ensuring the system's sustainability. Several techniques can be applied to achieve these goals, from model compression and local techniques to energy consumption control methods. In the integrated system we are discussing, on-device processing is one of the key positive features for reducing energy consumption, especially since the cost of transmitting data over the network far exceeds the cost of processing it locally. Another critical feature is Smart Sleep, where devices can enter an automatic or idle sleep mode when no abnormal activities are detected. As previously mentioned, the techniques of pruning, quantization, and compression can be employed to reduce model size, thereby helping to lower memory consumption. As discussed earlier, techniques such as pruning, quantization, and compression can be applied to shrink the model size, which in turn helps reduce memory usage. Hierarchical storage is another effective technique, ensuring that only critical data is recorded on the blockchain, with less important data stored in external databases or cached memory.

9.7   CHALLENGES AND FUTURE RESEARCH DIRECTIONS

Interoperability is one of the fundamental challenges in integrating TinyML and blockchain technologies due to the diversity of devices from different vendors and the lack of standardized protocols. In practice, ensuring compatibility between different systems is complex because of the differences in programming approaches and the underlying platforms used across devices. To overcome this challenge, establishing unified industry standards by academic and industrial organizations, like the IETF or IEEE, is necessary to guarantee seamless integration across various systems. Ensuring privacy in machine learning is critical in applications that handle sensitive data, such as those in healthcare or financial sectors. Using the federated learning model is an effective solution for privacy protection because of its ability to train without centralizing data. Instead, models are trained locally on edge devices, and only the modified models are transmitted to the central server.

Blockchain also strengthens privacy by providing secure, tamper-proof documentation for all updates made to the model. Smart contracts also ensure that transactions related to models are legal and secure. However, there remains a need to develop techniques to reduce the size of on-chain transactions in IoT systems that rely on resource-constrained devices.

In IoT environments, distributed networks encounter a major challenge in identifying suitable incentives for participants to promote cooperation among devices. Consensus mechanisms must encourage multiple parties to participate in the computational process or the shared storage of data. Some networks utilize traditional consensus mechanisms such as Proof of Work (PoW) or Proof of Stake (PoS), but these are unsuitable for environments with resource-constrained devices. To address this issue, lighter and more device-compatible consensus mechanisms, such as Delegated Proof of Stake (DPoS), emerged with lower power requirements and decreased network pressure. In the DPoS mechanism, not every node in the network takes part in verifying transactions, as is the case with Proof of Work (PoW). Instead, a limited number of trusted nodes, commonly referred to as delegates or validators, are elected to perform this task. These nodes are not technically disparate from the others but are temporarily charged with administering the process of consensus and verification, as voted on by the network stakeholders. This smart allocation of tasks enables consensus to be achieved more rapidly and efficiently without obliging every node to waste its resources in verification, thus being particularly well suited for IoT applications with limited energy and computational resources [17].

As the system was expanded to include more devices or cases, the demand for resources and the time required to process data increased. However, scaling may increase costs, particularly in storage or blockchain transaction operations. Therefore, researchers and developers must find solutions that balance scalability and cost. Approaches such as off-chain processing and hierarchical data storage can be employed to reduce the need to store all data in the distributed ledger. With the advancement of modern technologies, we are beginning to witness the emergence of innovations that will significantly impact the integration of TinyML and blockchain in Internet of Things environments. Neuromorphic chips are one such technology. In the context of TinyML, neuromorphic chips can greatly improve the speed and efficiency of data processing on edge devices since they emulate the human brain's information processing mechanism.

On the other hand, large language models at the edge (LLMs at the Edge), such as GPT, could enable edge devices to interact intelligently with users in IoT environments. By implementing natural language processing models at the edge, the devices can instantly analyze textual and spoken data without sending it to the cloud. Finally, directed acyclic graph (DAG)–based ledgers, such as IOTA, offer an innovative alternative to the traditional blockchain structure. Rather than sequentially organizing blocks, the DAG structure represents transactions concurrently, which enables faster and more flexible transactions with lower latency. Since these technologies do not require the intensive mining operations typical of traditional blockchain, they are particularly well suited for resource-constrained devices.

9.8 CONCLUSIONS AND REMARKS

Integrating TinyML technologies with blockchain in Internet of Things environments is crucial to enhancing security, efficiency, and privacy protection in smart applications. With the continuous rise in the adoption of smart devices in our daily lives, it becomes crucial to address security and data protection challenges. TinyML aims to provide artificial intelligence close to the user at the edge, enabling devices to perform local processing and reduce reliance on the cloud. Blockchain offers a security layer that ensures data integrity and authenticity. Integrating these technologies ensures integrity, reliability, and security. The architectural approach for this integration ensures a smooth data flow within the system. Starting with data collection from the environment through distributed sensors to local processing at the Edge layer using lightweight TinyML models. Then, a distributed ledger in the blockchain ensures integrity and prevents tampering. Data transmission across the network is managed using communication layer protocols. A case study was presented on an intrusion detection system implemented in smart homes, where TinyML was used to detect abnormal and unusual activities within the home. Alerts are sent from the edge devices to the blockchain network via communication protocols for future investigations. This approach ensures a secure and tamper-proof home environment. This approach still faces many challenges, particularly regarding interoperability, as the interconnected devices are diverse and have different versions. The lack of standards also presents a significant challenge. This is where the role of trusted industrial companies comes into play, as they must work on establishing suitable standards for this type of integration.

REFERENCES

1. L. S. Vailshery, "Number of IoT Connections Worldwide 2022–2033," *Statista*, Jun. 2024. [Online]. Available: https://www.statista.com/statistics/1183457/iot-connected-devices-worldwide/. [Accessed: May 16, 2025].
2. D. Rupanetti and N. Kaabouch, "Combining Edge Computing-Assisted Internet of Things Security with Artificial Intelligence: Applications, Challenges, and Opportunities," *Applied Sciences*, vol. 14, no. 16, p. 7104, 2024, doi: 10.3390/app14167104.
3. A. Elhanashi, P. Dini, S. Saponara, and Q. Zheng, "Advancements in TinyML: Applications, Limitations, and Impact on IoT Devices," *Electronics*, vol. 13, no. 17, p. 3562, 2024, doi: 10.3390/electronics13173562.

4. C. R. Banbury et al., "Benchmarking TinyML Systems: Challenges and Direction," in *Proceedings of the SysML Conference*, Austin, Texas, USA, 2020, doi: 10.48550/arXiv.2003.04821.

5. G. Habib, S. Sharma, S. Ibrahim, I. Ahmad, S. Qureshi, and M. Ishfaq, "Blockchain Technology: Benefits, Challenges, Applications, and Integration of Blockchain Technology with Cloud Computing," *Future Internet*, vol. 14, no. 11, p. 341, Nov. 2022, doi: 10.3390/fi14110341.

6. J. Byrum, "The Past, Present, and Future of the Payment System as Trusted Broker and the Implications for Banking," in Babich, V., Birge, J.R., Hilary, G. (eds). *Innovative Technology at the Interface of Finance and Operations*, Vol. 11, Springer, 2022, doi: 10.1007/978-3-030-75729-8_4.

7. K. Bandhu, R. Litoriya, P. Lowanshi, and M. Jindal, "Making Drug Supply Chain Secure, Traceable, and Efficient: A Blockchain and Smart Contract-Based Implementation," *Multimedia Tools and Applications*, vol. 82, no. 15, pp. 23541–23568, 2023, doi: 10.1007/s11042-022-14238-4.

8. P. Vionis and T. Kotsilieris, "The Potential of Blockchain Technology and Smart Contracts in the Energy Sector: A Review," *Applied Sciences*, vol. 14, no. 1, p. 253, 2024, doi: 10.3390/app14010253.

9. A. Carelli, A. Palmieri, A. Vilei, F. Castanier, and A. Vesco, "Enabling Secure Data Exchange through the IOTA Tangle for IoT Constrained Devices," *Sensors*, vol. 22, no. 4, p. 1384, 2022, doi: 10.3390/s22041384.

10. S. Heydari and Q. H. Mahmoud, "Tiny Machine Learning and On-Device Inference: A Survey of Applications, Challenges, and Future Directions," *Sensors*, vol. 25, no. 10, p. 3191, May 2025, doi: 10.3390/s25103191.

11. M. Zeba, M. Mashira, M. M. J. Mim, M. M. Hassan, et al., "Energy-Efficient Architecture for Optimized IoT Data Transmission from Edge to Cloud," Preprint, Mar. 2024, doi: 10.21203/rs.3.rs-4127989/v1.

12. Z. Huang, L. Yu, L. F. H. Contreras, and O. Kavehei, "Efficient and Secure μ-Training and μ-Fine-Tuning for TinyML Optimization and Personalization at the Edge," *medRxiv*, preprint, Feb. 2025, doi: 10.1101/2025.01.30.25321374.

13. N. N. Alajlan and D. M. Ibrahim, "DDD TinyML: A TinyML-Based Driver Drowsiness Detection Model Using Deep Learning," *Sensors*, vol. 23, no. 12, p. 5696, 2023, doi: 10.3390/s23125696.

14. E. E. Esther, P. S. Rani, and N. Kamal, "TinyML: A Cutting-Edge Technology Revolutionizing Machine Learning in Healthcare," *International Journal of Information and Communication Technology*, vol. 6, no. 1, pp. 24–28, 2025, doi: 10.33545/2707661X.2025.v6.i1a.107.

15. A. T. Khan, S. M. Jensen, and A. R. Khan, "Plant Disease Detection Model for Edge Computing Devices," *Frontiers in Plant Science*, vol. 14, p. 1308528, 2023, doi: 10.3389/fpls.2023.1308528.

16. P. Qi, D. Chiaro, F. Giampaolo, and F. Piccialli, "A Blockchain-Based Secure Internet of Medical Things Framework for Stress Detection," *Information Sciences*, vol. 628, pp. 377–390, May 2023, doi: 10.1016/j.ins.2023.01.123.

17. E. Ul Haque, A. Shah, J. Iqbal, S. S. Ullah, R. Alroobaea, and S. Hussain, "A Scalable Blockchain-Based Framework for Efficient IoT Data Management Using Lightweight Consensus," *Scientific Reports*, vol. 14, p. 7841, 2024, doi: 10.1038/s41598-024-58578-7.

18. J. M. Fartitchou, I. Lamaakal, Y. Maleh, K. El Makkaoui, Z. El Allali, P. Pławiak, F. Alblehai, and A. A. Abd El-Latif, "IOTASDN: IOTA 2.0 Smart Contracts for Securing Software-Defined Networking Ecosystem," *Sensors*, vol. 24, no. 17, pp. 1–25, Sep. 2024, doi: 10.3390/s24175716.

19. LIBM, "KAYA&KATO and IBM Pioneer Blockchain Network to Track Sustainable Clothing," *IBM Newsroom*, Nov. 16, 2020.

20. Z. A. Khan, N. Javaid, A. Saeed, I. Ahmed, and F. A. Khan, "Towards IoT device privacy & data integrity through decentralized storage with blockchain and predicting malicious entities by stacked machine learning," *Internet of Things*p. 101642, 2025.

21. F. Khan, A. A. Al-Atawi, A. Alomari, A. Alsirhani, M. M. Alshahrani, J. Khan, and Y. Lee, "Development of a Model for Spoofing Attacks in Internet of Things," *Mathematics*, vol. 10, no. 19, pp. 1–16, 2022, doi: 10.3390/math10193686.

22. J. Atetedaye, "Cybersecurity in the Internet of Things (IoT): Challenges and Solutions," *International Journal of Scientific Research in Modern Science and Technology*, vol. 3, no. 7, pp. 17–21, Jul. 2024, doi: 10.59828/ijsrmst.v3i7.222.

23. C. Fathy and H. M. Ali, "A Secure IoT-Based Irrigation System for Precision Agriculture Using the Expeditious Cipher," *Sensors*, vol. 23, no. 4, p. 2091, 2023, doi: 10.3390/s23042091.

24. S. A. R. Zaidi, A. M. Hayajneh, M. Hafeez, and Q. Z. Ahmed, "Unlocking Edge Intelligence Through Tiny Machine Learning (TinyML)," *IEEE Access*, vol. 10, pp. 100867–100877, 2022, doi: 10.1109/ACCESS.2022.3207200.

25. T. Suwannaphong, F. Jovan, I. Craddock, and R. McConville, "Optimising TinyML with Quantization and Distillation of Transformer and Mamba Models for Indoor Localization on Edge Devices," *Scientific Reports*, vol. 15, no. 1, Art. no. 10081, Mar. 2025, doi: 10.1038/s41598-025-94205-9.

26. Y. Chen, Y. Qiu, Z. Tang, S. Long, L. Zhao, and Z. Tang, "Exploring the Synergy of Blockchain, IoT, and Edge Computing in Smart Traffic Management across Urban Landscapes," *Journal of Grid Computing*, vol. 22, no. 1, p. 45, Apr. 2024, doi: 10.1007/s10723-024-09762-6.

27. A. Aljumah, "Blockchain-Inspired Distributed Security Framework for Internet of Things," Scientific Reports, vol. 15, Article no. 10066, Mar. 2025, doi: 10.1038/s41598-025-93690-2.

28. K M. Shalan, M. R. Hasan, Y. Bai, and J. Li, "Enhancing Smart Home Security: Blockchain-Enabled Federated Learning with Knowledge Distillation for Intrusion Detection," *Smart Cities*, vol. 8, no. 1, p. 35, Feb. 2025, doi: 10.3390/smartcities8010035.

29. V. Tsoukas, A. Gkogkidis, A. Kampa, G. Spathoulas, and A. Kakarountas, "Enhancing Food Supply Chain Security Through the Use of Blockchain and TinyML," *Information*, vol. 13, no. 2, p. 213, 2022, doi: 10.3390/info13050213.

30. S. Adhikary, S. Dutta, and A. D. Dwivedi, "TinyWolf—Efficient on-device TinyML training for IoT using enhanced Grey Wolf Optimization," *Internet of Things*, vol. 28, p. 101365, 2024, doi: 10.1016/j.iot.2024.101365.

31. M. Banbury et al., "Micronets: Neural Network Architectures for Deploying TinyML Applications on Commodity Microcontrollers," in *Proceedings of the IEEE/CVF Conference on Computer Vision and Pattern Recognition Workshops (CVPRW)*, 2021, pp. 1420–1429, doi: 10.1109/CVPRW53098.2021.00160.

32. S. Wang, H. Li, J. Chen, J. Wang, and Y. Deng, "DAG Blockchain-Based Lightweight Authentication and Authorization Scheme for IoT Devices," *Journal of Information Security and Applications*, vol. 66, p. 103134, Mar. 2022, doi: 10.1016/j.jisa.2022.103134.

33. M. Smyrlis, E. Floros, I. Basdekis, and G. Spanoudakis, "RAMA: A Risk Assessment Solution for Healthcare Organizations," *Computers in Biology and Medicine*, vol. 152, p. 106400, 2023, doi: 10.1016/j.compbiomed.2023.106400.

34. L H. Kayan, Y. Majib, W. Alsafery, M. Barhamgi, and C. Perera, "An Adaptable and Unsupervised TinyML Anomaly Detection System for Extreme Industrial Environments," *Sensors*, vol. 23, no. 4, pp. 2344, 2023, doi: 10.3390/s23042344.

35. J. H. Lee, S. H. Kim, and H. J. Kim, "Blockchain-Based Community Safety Security System with IoT Integration," *Sustainability*, vol. 13, no. 24, p. 13994, Dec. 2021, doi: 10.3390/su132413994.

36. Rashid, A., Masood, A., and Khan, A. U., "ACS-IoT: Smart Contract and Blockchain Assisted Framework for Access Control Systems in IoT Enterprise Environment," *Wireless Personal Communications*, vol. 136, pp. 1331–1352, 2024, doi: 10.1007/s11277-024-11266-1.

37. K. S. Qader and K. Cek, "Influence of Blockchain and Artificial Intelligence on Audit Quality: Evidence from Turkey," *Heliyon*, vol. 10, no. 9, p. e30166, Apr. 2024, doi: 10.1016/j.heliyon.2024.e30166.

38. M. Abbasi, J. Prieto, M. Plaza-Hernández, and J. M. Corchado, "Proof-of-Resource: A Resource-Efficient Consensus Mechanism for IoT Devices in Blockchain Networks," *EAI Endorsed Transactions on Internet of Things*, Nashville, TN, USA, vol. 10, 2024, doi: 10.4108/eetiot.6565.

39. W. Villegas-Ch, R. Gutierrez, A. Maldonado Navarro, and A. Mera-Navarrete, "Lightweight Blockchain for Authentication and Authorization in Resource-Constrained IoT Networks," *IEEE Access*, vol. 13, pp. 1792–1809, Jan. 2025, doi: 10.1109/ACCESS.2025.3551261.

40. G. Goint, C. Bertelle, and C. Duvallet, "Secure Access Control to Data in Off-Chain Storage in Blockchain-Based Consent Systems," *Mathematics*, vol. 11, no. 7, pp. 1–18, 2023, doi: 10.3390/math11071592.

41. E. Rhee and J. Lee, "Improved DAG in Blockchain Tangle for IOTA," *Indonesian Journal of Electrical Engineering and Computer Science*, vol. 34, no. 2, pp. 806–813, May 2024, doi: 10.11591/ijeecs.v34.i2.pp806-813.

10 TinyML
Emerging Applications and Future Research Directions

Chaymae Yahyati, Ismail Lamaakal,
Khalid El Makkaoui, Ibrahim Ouahbi,
and Yassine Maleh

10.1 INTRODUCTION

The pursuit of a better quality of life, combined with the growing array of global challenges, is driving rapid technological advancement. "Edge Intelligence" (EI) emerges as a promising solution by integrating the capabilities of the Internet of Things (IoT) and machine learning to tackle problems directly at the data source. Intelligent cameras with on-device processing and network capabilities are increasingly being adopted in embedded vision systems, finding applications in both industrial and household environments. At any given time, IoT sensor networks collect gigabytes of data, requiring immediate analysis through onboard computing resources. This real-time processing enables the extraction of meaningful, actionable insights for both individuals and interconnected systems [1]. Tasks that once demanded human expertise and complex hardware/software configurations are now being streamlined and automated using optimized machine learning models deployed on efficient embedded IoT platforms.

Within the domain of Edge Intelligence (EI), the central emphasis is placed on sensory systems, including camera-based configurations, audio sensors, and features like traffic monitoring in smart cities. EI essentially functions as a broad sensory network, continuously observing and analyzing environmental events. In an integrated technological framework, the collected data may also be transmitted to the cloud for deeper analysis.

Over the past decade, notable progress has been made in software engineering for machine learning and embedded IoT systems, helping developers overcome many of the challenges associated with deploying deep learning models [2,3,4,5]. EI is particularly well-suited for situations demanding rapid decision-making and real-time responses to critical data streams. Edge AI application areas can be distinguished based on constraints such as power efficiency, physical size, and processing load. Among these, power efficiency is especially crucial, as edge devices typically operate under low-power conditions—common in smartphones, wearables,

DOI: 10.1201/9781003544449-10

and IoT environments. As a result, AI models must be carefully optimized to minimize energy consumption, ensuring longer battery life and extended device usability.

Moreover, the compact nature of edge devices requires that AI models be both lightweight and spatially efficient. This becomes especially important in applications involving drones, robots, or wearable devices, where the size and weight directly influence performance, mobility, and user comfort.

10.2 OVERVIEW OF TinyML

This section outlines the software engineering workflow for designing and deploying TinyML models [6]. The development process is typically divided into two main stages: the first takes place in a training environment using high-performance computing resources such as CPUs or GPUs, while the second focuses on deployment using resource-constrained edge hardware for real-world applications, as illustrated in Figure 10.1.

At the outset, customer or business requirements are broken down into three interrelated categories:

1. **Model Requirements:** Specifying the expected performance, accuracy, and complexity of the machine learning model.
2. **Data Requirements:** Defining the type, volume, and quality of data needed for effective training and inference.
3. **System Requirements:** Detailing hardware constraints, latency tolerances, power consumption, and integration needs for the target edge platform.

This methodology enables the delegation of specific responsibilities to distinct teams or entities, each accountable for delivering particular components. It also empowers these teams with a degree of autonomy, allowing them to thoroughly understand the requirements and apply their expertise in areas such as system architecture, model development, and the creation of data pipelines for collection, cleaning, and labeling.

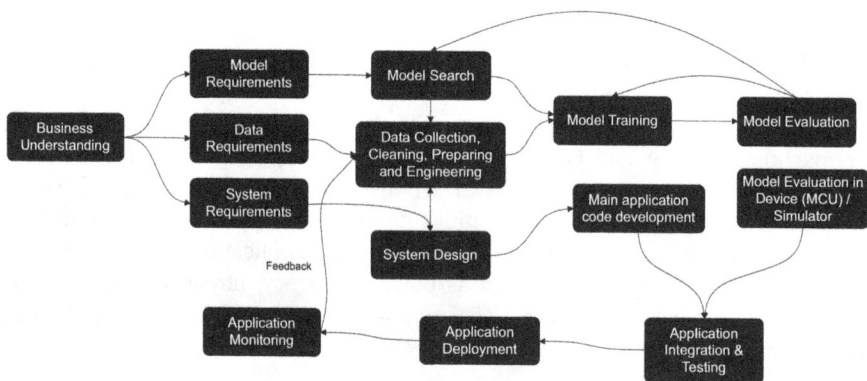

FIGURE 10.1   Development and deployment workflows for a TinyML model.

The execution of the process flow proceeds as follows:

- **System Design and Constraints:** Once the system architecture is defined, it provides key insights that guide both model exploration and data engineering tasks. Given that the application is intended to run on a compact microcontroller unit (MCU), the system design imposes constraints on the computational and memory requirements of the algorithms used for data preprocessing, postprocessing, and machine learning. Importantly, application development can begin concurrently with model development.
- **Parallel Development:** This parallel workflow allows embedded developers to design and debug heuristic pipelines for sensing and actuation, based on expected outputs from the machine learning model. Developers can also simulate model outputs using calibration functions, enabling them to fine-tune their embedded code even before the final model is complete.
- **Model Training and Evaluation:** Adhering to established best practices in model training, developers can identify a suitable model and assess its performance on a validation dataset. Insights from this evaluation phase guide decisions regarding further training or hyperparameter tuning to improve model performance.
- **Model Compilation:** Once the model meets the desired validation criteria, it proceeds to the compilation stage, managed by the TensorFlow Lite Micro (TFLM) framework. This step involves optimizing the model using hardware-specific libraries that leverage the custom features of the target MCU.
- **Post-Compilation Testing:** After compilation, the model is evaluated either within a simulator or directly on the target device, depending on resource availability. If there is a significant drop in performance post-compilation, the model should be sent back for necessary adjustments during the compilation phase.
- **Integration and Deployment:** Once the compiled model exhibits performance consistent with the original model, it is handed over to the application integration team. This team embeds the model into the application, performs testing with a designated test dataset, and subsequently deploys it in real-world environments.
- **Monitoring and Continuous Improvement:** Collecting performance data from real-world usage provides valuable feedback for continuous integration. These insights help identify performance gaps or highlight the need to retrain the model using new data from evolving real-world distributions.

10.3 WORKFLOW STEPS, CHALLENGES, AND PROPOSED SOLUTIONS

Currently, there is limited understanding of the most effective and efficient methodologies for developing, deploying, and maintaining IoT-based embedded vision systems tailored to TinyML applications. A notable gap exists in the availability of well-established agile development practices specifically designed for the unique requirements of TinyML engineering. Building such applications necessitates inter-disciplinary expertise in embedded systems, computer vision, and machine learning. However, given the relatively nascent stage of the field, there is a shortage of engineers with the comprehensive skill set required to address the multifaceted challenges of TinyML across diverse application domains.

The steps involved in developing and implementing TinyML applications and the associated challenges and proposed solutions are as follows:

1. **Business (Customer) Understanding:**
   - **Challenges:**
   - Customer requirements are evolving rapidly.
   - TinyML engineering demands proficiency across various domains, including AI, machine learning, and embedded systems. SMEs/business users may not have the proficiency to understand the inputs required for these systems.
   - **Solution:**
   - Separating teams responsible for system creation, data handling, and model development facilitates concurrent work [7].
   - Optimizing the TinyML engineering process, version control is essential to meet customer needs. This allows for incremental enhancements on specific features demanded by the consumer [8].

2. **System Requirements:**
   - **Challenges:**
     - Deploying in real-time environs poses significant challenges.
     - Ensuring compatibility between hardware systems and the mobility of models around various hardware configurations is necessary.
     - Addressing performance degradation in real-time scenarios is a primary concern for the edge deployment of TinyML systems.
     - The absence of standardized tools for comparing performance across different algorithms and MCU systems is a notable issue [9].
   - **Solution:**
     - Establishing shared implicit assumptions concerning visual data, such as illumination conditions and shadows.
     - Managing the exponential increase in hardware device variations by strictly controlling the types of devices and chipsets utilized in TinyML systems.
     - Making design choices based on the environmental obstacles anticipated by the TinyML application.
     - Augmenting the training dataset with data is crucial for this purpose [10].

3. **Model Requirements:**
   - **Challenges:**
     - Assessing the impacts of different levels of quantization and precision, along with model peak memory requirements, presents significant challenges.

- Documentation regarding the selection of models suitable for TinyML projects is lacking.
- Identifying practical models for resource-constrained edge environments is a complex endeavor.
- **Solution:**
  - TinyML systems must choose target MCU hardware for their projects according to requirements such as model size, inference speed, and power constraints specific to the application [10].
  - Constrained neural architecture search (NAS) can help narrow down model architectures that are relevant to the system design choices [11,12].

4. **System Design:**
   - **Challenges:**
     - At present, there is no clear-cut framework or standard benchmark to aid developers of TinyML-powered IoT-embedded vision applications in determining the most suitable hardware and model designs that align with particular application, market, or customer needs.
     - Maintaining model development tools for distinct target hardware systems can be laborious.
     - Model development becomes highly complex while addressing a wide range of applications.
     - Vision data from videos frequently includes temporally redundant information, resulting in added unnecessary computing on resource-constrained edge hardware.
   - **Solution:**
     - Ensuring the correct setup of the TinyML TFLM framework infrastructure from the outset will facilitate faster iteration [13].
     - Ensuring efficient management of models and maintaining version control are essential for optimizing TinyML development efforts.
     - Utilizing event-driven methods for visual data processing can reduce the need for repetitive computations [14].

5. **Data Engineering:**
   - **Challenges:**
     - Curated datasets extracted from IoT-embedded vision sensors for various applications are lacking.
     - Model data obtained from various sensors and sources may come in different file formats.
     - Datasets may not adequately represent real-world conditions found in distributions.
   - **Solution:**
     - Access to data is only available through programmatic interfaces.
     - Data augmentation involves emulating real-world scenarios by introducing variations like scaling, noise, shadows, color, and lighting conditions.

6. **Model Training:**
- **Challenges:**
  - Basic joint optimization methods such as NAS, quantization, and pruning might interact in manners that result in less than optimal outcomes [15].
  - The absence of emulation tools for TinyML engineering leads to increased time and effort required for model development [16].
  - The hyperparameter search process may encounter challenges in reproducing results.
  - Developing TinyML models with a performance-oriented approach presents challenges.
  - Automation of TinyML engineering via web-based tools.
- **Solution:**
  - Using deterministic randomness, such as exposed random seeds, assists in iterating during hyperparameter tuning.
  - TinyML application developers ought to leverage AutoML tools to automate the training process, especially when confronted with abundant training and validation data.
  - Startups like Edge Impulse and Qeexo provide AutoML capabilities accessible over the web, which can be directly integrated into edge applications.

7. **Model Evaluation:**
- **Challenges:**
  - TinyML models must demonstrate robustness to handle environmental anomalies and noise effectively.
  - There is a dearth of uniform TinyML evaluation metrics.
- **Solution:**
  - Test-driven development should focus on assessing both local and global adversarial robustness [17].
  - ML Commons has introduced the ML PerfTiny benchmark, establishing the first industry standard benchmark suite for ultralow-power machine learning [18].

8. **Model Compilation:**
- **Challenges:**
  - Since MCU designs vary widely, it is advantageous to utilize a framework with general compiler capabilities to ensure design flexibility across different platforms. However, such a generic compiler may not fully exploit the distinct hardware features present in certain MCU families.
  - Ensuring the security of embedded devices poses a significant challenge in IoT applications [19,20].
  - IoT-embedded vision devices should minimize communication to conserve power consumption.
  - IoT-embedded devices exhibit significant heterogeneity. While manual coding and code generation can harness tailored optimizations, they may sacrifice flexibility in the process.

- Compiler configurations should be fine-tuned to match the hardware system targeted by the chosen system design options.
- Implementing device training poses challenges for devices with limited memory footprint [21].
- Assessing performance is only possible after deploying an application onto the edge device system.

- **Solution:**
  - Compilers should offer admittance to performance metrics for MCUs due to their single-threaded nature.
  - A hardware-aware compiler can automatically make advantageous trade-offs without requiring deep knowledge of MCU architectures [22,23].

9. **Compiled Model Evaluation:**
  - **Challenges:**
    - Debugging of compiled models on edge devices is challenging due to computational limitations.
    - Limitations in generic frameworks hinder hardware awareness of the TinyML workflow.
  - **Solution:**
    - TinyML development requires the use of device simulators and emulators.
    - Device simulators can help to compare the performance of raw models and different versions of compiled TinyML models.

10. **Application Integration:**
  - **Challenges:**
    - Given that edge applications often require time-sensitive decisions, ensuring that the reliability of TinyML systems is paramount. It is essential to embed robustness into TinyML models to enable secure fail-over mechanisms in uncertain environments [20].
    - Power profiling is complicated because the data pathways and preprocessing phases can differ substantially between devices. TinyML-capable devices exhibit vastly different power consumption, complicating the task of maintaining consistent accuracy across them.
    - Real-world circumstances can rapidly evolve because of changes in the sensing environment or the senescence of vision sensors. TinyML addresses a substantial challenge in online learning by offering a solution despite the limited computing resources available.
    - Optimizing design trade-offs is challenging until the model itself is optimized.
  - **Solution:**
    - Automated tools can help explore trade-offs between performance, precision, and power [24].

11. **Application Testing:**
  - **Challenges:**
    - Debugging application issues becomes more challenging as the system encompasses heuristic code alongside trained TinyML models.

- TinyML-powered applications are streamlined, lacking robust tools for data measurement and visualization. This presents supplementary hurdles for debugging throughout model development and deployment because of memory limitations [21].
- Testing TinyML models can be wrecked by biases in the tests.
- **Solution:**
  - Examining bug taxonomy for IoT applications can help preempt critical issues in application code [25,26].
  - Having devoted quality assessment squads can facilitate fair debugging by mitigating developer bias.

12. **Application Deployment:**
  - **Challenges:**
    - Models installed on TinyML systems should support upgrades with minimal intervention or system downtime.
    - Deploying TinyML models in practical settings continues to be difficult despite successful model validation during training.
    - Ensuring that applications can run seamlessly across different devices and vendors presents a significant obstacle in TinyML engineering.
    - Rectifying bugs in embedded application deployment presents significant challenges.
  - **Solution:**
    - Achieving reproducibility relies on the distinct identification and accessibility of algorithms, parameters, and data/labels in real-world settings [10].
    - Having over-the-air updates is crucial for seamlessly deploying bug fixes and upgrades for both firmware and TinyML models.

13. **Application Monitoring:**
  - **Challenges:**
    - Enhancing performance in real-world settings poses challenges due to constraints on edge hardware.
    - Measuring the performance of TinyML applications in real-time scenarios is difficult.
  - **Solution:**
    - Incorporation of labeling and retraining into continuous integration processes, informed by feedback from surveillance, can enhance application performance by adapting to new data distributions confronted in real-time scenarios.
    - Utilizing device-level surveillance tools can enhance debuggability in real-world deployments.

10.4   APPLICATIONS OF TinyML AND FUTURE RESEARCH DIRECTIONS

The rapid growth of the TinyML movement is largely driven by its diverse range of applications, which not only generate valuable datasets but also accelerate ongoing research. This, in turn, creates a continuous demand for more efficient TinyML systems and iterative improvements in design methodologies.

Initially centered around vision tasks, TinyML applications have, in recent years, significantly broadened their scope to encompass natural language processing, predictive modeling, pattern recognition and classification, and data analysis. This section provides a concise overview of these application domains, along with a high-level evaluation of the key challenges and future opportunities.

At present, TinyML has demonstrated its effectiveness across a wide array of use cases, including smart devices (from individual sensors to entire urban infrastructures), industrial monitoring and control, healthcare diagnostics, security and surveillance, administrative automation, and financial operations. These applications have a meaningful and growing impact on daily life and societal systems [27,28–32].

10.4.1 SMART VEHICLES

An intelligent vehicle is characterized by its integrated computing, storage, and communication capabilities, which enable it to learn from its environment and make informed decisions. These vehicles are equipped with internal and external sensors, as well as multi-interface communication modules. The growing adoption of smart vehicles—featuring onboard wireless devices and advanced sensors such as radar and LiDAR—has shifted attention toward efficient traffic management and transportation systems aimed at minimizing congestion and reducing travel times.

Smart vehicles offer a variety of advanced features, including real-time information exchange and precise location tracking. These capabilities support a wide range of specialized applications, particularly in safety communication and early warning systems. Within a vehicular edge computing (VEC) framework, vehicles are typically equipped with onboard units (OBUs) that facilitate wireless communication. In emergency scenarios, such as accident detection, onboard sensors play a pivotal role—for instance, in determining whether airbags have been deployed, which can trigger immediate disaster response protocols.

10.4.1.1 Smart Vehicle Services

- **Assisted Driving:** In modern transportation systems, vehicles such as cars, buses, and trains are increasingly equipped with the ability to transmit critical information—ranging from accident reports and road closures to real-time traffic congestion updates. This capability is made possible through the integration of sensors, actuators, and onboard processors, which collectively enhance both safety and navigation. The traffic pattern data generated by these intelligent systems can be highly beneficial to a wide range of stakeholders, including public agencies, private enterprises, and emergency services [33]. According to the National Highway Traffic Safety Administration (NHTSA), intelligent vehicles are classified into five distinct levels, each representing a progressive degree of automation and autonomy [34].
- **Autonomous Vehicles:** As smart vehicles advance toward full autonomy, establishing reliable and high-speed connectivity among them becomes essential. The emergence of vehicular networks—driven by this technological evolution—is central to the development of intelligent transportation systems and the broader vision of smart cities. These networks are expected

to support a wide range of advanced applications, including enhanced road safety, improved traffic flow, autonomous driving capabilities, and uninterrupted access to Internet-based services [35,36].

The global momentum toward automated vehicles is reshaping the automotive industry. However, realizing fully autonomous vehicles still faces significant challenges related to security, reliability, and privacy. One of the most pressing concerns is the vulnerability of autonomous vehicles to cyberattacks. A single malicious intrusion into a vehicle's software system could result in cascading failures and potentially severe accidents. Moreover, the interconnected nature of these systems exposes them to unauthorized access and unpredictable cyber threats.

To address safety-critical concerns, autonomous vehicle designs incorporate mechanisms that allow the vehicle to detect and react to potential hazards while continuously monitoring environmental conditions throughout the journey. These systems typically assume that the driver will input the destination or route; however, they do not rely on continuous human supervision, underscoring the importance of robust automation to ensure safe operation [37]. Although automated vehicular systems and connected vehicular technologies are distinct, they share overlapping features and often operate in complementary ways within the broader intelligent transportation ecosystem.

- **Smart Parking:** In metropolitan areas, a considerable number of vehicles remain parked across diverse locations, including street-side and off-street parking facilities. Unlike vehicles in motion, parked vehicles remain stationary for extended durations. Although they do not relay information through movement, those equipped with wireless communication modules and rechargeable batteries—as part of the smart street vehicle (SSV) paradigm—can serve as valuable components of vehicular communication infrastructure. These parked SSVs possess unique characteristics, enabling them to communicate with each other and connect with nearby moving SSVs. Acting as static communication nodes, they help form a resilient backbone that enhances overall vehicular network connectivity.

  The effectiveness of parked vehicles as communication infrastructure is influenced by factors such as the number of vehicles present in a given area and their average parking duration [28]. Collaborative networking among parked SSVs, especially in densely occupied parking lots, facilitates the execution of resource-intensive computational tasks under stable communication conditions. Individual vehicles, often limited in computational and energy resources, may struggle with such demanding tasks. Parked SSVs mitigate this limitation by offering idle yet capable processing power, enabling efficient task completion in a shorter time frame. In essence, these clusters of parked smart vehicles function as localized microdata centers, capable of handling complex and high-computation workloads in a distributed and energy-efficient manner.

10.4.1.2   Smart Vehicle Applications

- **Safety Applications:** A key objective of intelligent transportation applications is to enhance safety by reducing the risk of accidents. These systems continuously monitor the driving environment and issue alerts to drivers when potentially dangerous situations are detected. One example involves the deployment of a global camera sensor installed at traffic monitoring points, capable of detecting motion within its field of view and recognizing vehicle license plates. Upon detection, the sensor logs both the vehicle's location and its identification number, transmitting this data to a local edge server for further processing.

  In parallel, a smart local camera sensor (LCS) mounted at the front of a vehicle monitors the driver's behavior. The LCS is designed to issue real-time warning messages when it detects risky activities, thereby contributing to accident prevention. Timely and repeated alerts enhance driver awareness and help mitigate unsafe driving practices. Additionally, the LCS is capable of broadcasting a unique vehicle identifier to the edge server, along with evidence of the detected activity. This enables the server to maintain records of disruptive or hazardous behaviors [38–41].

  Furthermore, context-aware systems are increasingly employed to improve the adaptability and intelligence of vehicular applications. These systems utilize user-specific and environmental data to dynamically adjust their functionality. Context-aware applications operate by acquiring, processing, and acting upon contextual information [42]. By leveraging contextual awareness, these systems can generate succinct and relevant messages, thereby reducing the need for extensive radio communication. Users can then extract the desired content using advanced decoding techniques and big data analytics, including natural language processing and other AI-driven tools [39,43,44,45].

- **Non-Safety Applications:** The scope of vehicular edge computing (VEC) extends beyond safety-critical applications to encompass a wide array of non-safety services, including multimedia applications such as video streaming, augmented reality (AR), and infotainment. The growing popularity of streaming-based services has significantly increased network traffic, especially within the Internet of Things (IoT) ecosystem, where video content transmission constitutes a substantial portion [46]. This trend is particularly evident in mobile applications such as video crowdsourcing, where users contribute real-time footage via smartphones [47].

  Within the framework of the Internet of Vehicles (IoV), users connect mobile devices to remote servers to access multimedia content in transit, supporting applications in intelligent transportation systems and mobile entertainment. However, delivering high-quality streaming in IoV environments presents challenges due to factors such as jitter, buffering, throughput limitations, and latency—issues exacerbated by the high mobility of vehicles.

To address these challenges, a distributed and reliable real-time streaming scheme in vehicular cloud-fog networks is proposed in Ref. [48]. This approach employs a utility function to optimize Quality of Service (QoS) and ensure fairness in resource allocation among mobile users. The mechanism considers content availability and token-based reservation across service providers, edge nodes, and cloud infrastructure. Mobile devices forecast their future positions, estimate streaming data needs, and request the necessary tokens, enabling efficient content provisioning and enhancing the reliability of streaming utility.

For parking lot monitoring, Ref. [49] introduces an edge computing-based solution in which each vehicle uploads street-level video data collected by onboard cameras. The "ParkMaster" system uses this information—alongside GPS and inertial sensor data—to accurately detect and track parked vehicles in real time.

Augmented reality (AR) is another emerging multimedia application within VEC, offering a hybrid view that superimposes virtual content onto the real world. AR enhances situational awareness for drivers and pedestrians alike. A key implementation is the Head-Up Display (HUD), which projects AR-based navigation and safety information onto the windshield, minimizing driver distraction and improving road safety. A detailed investigation into HUD-based AR navigation systems is provided in Ref. [50], illustrating their potential for both safety and convenience services [51].

One innovative use case is AR-enabled walk navigation, which combines GPS and camera inputs to provide real-time, context-aware directions. Edge computing processes the live camera feed and overlays virtual navigation cues onto real-world objects, offering seamless guidance without compromising safety. Given the intensive storage, processing, and low-latency demands of AR, VEC emerges as the ideal framework for supporting AR applications in vehicular networks, leveraging mobility awareness, location specificity, and computational proximity.

10.4.2 Anomaly Detection

Anomaly detection, also known as outlier detection, involves identifying patterns or data points that deviate from expected behavior. This capability is essential in various IoT-based smart systems, including smart cities, environmental monitoring, and energy management applications. In such systems, sensors serve as primary input devices, continuously generating large volumes of data that are transmitted to cloud servers for analysis, decision-making, and long-term storage [52].

The literature outlines a broad spectrum of anomaly detection techniques, encompassing geometrical models, statistical approaches, and machine learning-based methods [53,54,55]. In recent years, there has been a marked shift toward data-driven strategies, with increased reliance on machine learning and deep learning to identify abnormal events [56]. These approaches typically involve training models on large-scale sensor datasets to differentiate between normal and anomalous conditions.

Cloud-based architectures are widely adopted in anomaly detection systems, particularly condition monitoring applications [57,58]. For example, in industrial machine health monitoring, edge devices collect vibration signals and send them to cloud servers, where machine learning algorithms detect irregularities indicative of potential faults. Despite the effectiveness of such cloud-centric frameworks, they are not without limitations. Challenges such as latency, dependency on constant connectivity, and data privacy concerns remain critical issues.

TinyML offers a transformative alternative by enabling machine learning inference directly at the edge. This paradigm facilitates real-time anomaly detection by eliminating the need to offload data to remote cloud servers. As a result, TinyML supports low-latency, energy-efficient, and privacy-preserving anomaly detection, empowering edge devices to autonomously identify irregular patterns and take immediate action.

10.4.2.1   Condition Monitoring

A practical implementation of a TinyML model on the STMicroelectronics STM32H743Z12 microcontroller unit (MCU) demonstrates the feasibility of anomaly detection in rotating machinery. In this setup, the MCU interfaces with an accelerometer to capture vibration signals, processes the data to extract relevant features, and stores the processed data locally. An autoencoder-based machine learning model is then trained directly on the MCU to detect deviations indicative of mechanical faults.

In another application, anomaly detection in submersible pumps used in wastewater treatment plants is achieved through a retrofitting kit integrated with a low-power MCU [59]. This kit comprises temperature and vibration sensors, an ESP32DEVKIT MCU, and power line communication components, all installed within the pump terminal chamber. The MCU collects sensory data, performs feature extraction, and uses a locally trained isolation forest model to detect anomalies in the data stream. Following the training phase, the system enters inference mode, where the MCU continues to perform real-time monitoring and dynamically updates the model as new data becomes available [60].

10.4.2.2   Predictive Maintenance

A TinyML model deployed on the ESP-WROOM-32 microcontroller unit (MCU) is used for anomaly detection in thermal imaging applications [61]. This device utilizes a convolutional neural network (CNN) developed with Keras and converted for embedded deployment using the tinymlgen library from Eloquent Arduino. The anomalies detected by the model are communicated to a remote server via the lightweight message queuing telemetry transport protocol (MQTT), with transmissions triggered only upon the detection of irregularities, thus conserving communication resources.

In the domain of water distribution systems, an online learning model named the Deep Echo State Network (DeepESN), a variant of recurrent neural networks (RNNs), has been proposed for real-time anomaly detection. This model is optimized for deployment in MCUs, offering a practical solution to assess reliability in distributed water infrastructures [62].

Furthermore, for detecting oil leaks in wind turbines, a Block-Based Binary Shallow Echo State Network (BBS-ESN) has been introduced as a quantized anomaly detection model [63]. This approach combines DeepESN with binarized input images and applies 1-bit quantization to both weights and activation functions, significantly reducing computational complexity. The performance of this model has been evaluated on the STMicroelectronics NUCLEO-H743ZI board, demonstrating its suitability for resource-constrained environments.

10.4.2.3   Internet of Intelligent Vehicles

An MCU-based system has been integrated into a vehicle to detect road surface anomalies, including potholes, speed bumps, and obstacles. The implementation employs the Arduino Nano 33 IoT board, which collects data from an onboard accelerometer and executes an unsupervised TinyML algorithm known as TEDA (Typicality and Eccentricity Data Analytics). Real-world validation experiments were conducted on asphalt pavement to assess the system's performance and reliability in detecting various types of road irregularities [64].

10.4.3   SMART FARMING

The rapid growth of the global population is driving an increased demand for agricultural output to ensure food security in the future. However, this rising demand is met with mounting challenges, including climate change, soil degradation, resource depletion, and population pressure. These issues necessitate the adoption of innovative agricultural methods. In response, farmers, researchers, and agricultural industries are increasingly turning to advanced technologies such as the Internet of Things (IoT), unmanned aerial vehicles (UAVs), artificial intelligence (AI), big data analytics, cloud computing, fog computing, and edge computing. These innovations are transforming traditional agriculture into intelligent, sustainable, and efficient systems—collectively referred to as smart agriculture or smart farming [65].

Among these technologies, IoT plays a central role by enabling continuous monitoring of soil and crop health, disease detection, and drone-based tracking of plant growth. Simultaneously, advancements in electronics have led to the development of high-performance, cost-effective components, such as microcontroller units (MCUs), sensor modules, and wireless transceivers. Modern MCUs now support not only basic sensing and control functions but also more complex tasks like executing machine learning models. Furthermore, advances in radio communication have enabled long-range data transmission with minimal energy consumption.

In a typical smart farming IoT architecture, sensor-equipped devices monitor variables such as soil moisture, crop health, irrigation status, greenhouse conditions, and weather. Data are collected via cameras and environmental sensors and transmitted through wireless sensor networks (WSNs) to cloud platforms. Cloud-based analysis provides farmers with actionable insights, including early detection of plant diseases and correlations between soil, weather, and crop performance, facilitating data-driven decision-making [66]. Machine learning is vital in this context, uncovering patterns in the data and forming the backbone of intelligent decision support systems.

Depending on the application, smart farming architectures are commonly organized into two-layer (physical–edge), three-layer (physical–edge–cloud), or four-layer (physical–edge–fog–cloud) structures, with the three-layer model being most prevalent [67].

- The **physical layer** consists of drones, sensors, and actuators that collect data from the field, animals, greenhouses, and atmospheric conditions.
- The **edge layer** includes low- and medium-capability computing devices that analyze and preprocess this data close to the source.
- The **cloud layer** acts as the central data repository and computational hub, capable of training and deploying large-scale machine learning models.

While cloud computing offers the necessary computational power for executing sophisticated ML models with millions of parameters, it introduces challenges in regions with limited internet connectivity, such as parts of Africa [68]. Issues such as latency, data transmission delays, and privacy concerns are significant drawbacks. Moreover, reliance on cloud infrastructure increases vulnerability to cybersecurity threats [69,70].

To address these limitations, computing paradigms such as fog and edge computing have emerged, bringing data processing closer to the devices that generate it. A particularly promising advancement in this domain is TinyML, which enables sensor devices to run machine learning inference locally. This capability supports tasks such as crop health monitoring, disease detection, and yield prediction without dependence on cloud connectivity. TinyML offers significant advantages in latency reduction, energy efficiency, privacy, and security [67].

TinyML holds immense promise for enhancing agricultural practices, especially in under-resourced regions like Africa, where digital agriculture and AI adoption remain limited. A noteworthy initiative in this direction is the PlantVillage project by Penn State University [71], which developed the Nuru mobile application. Nuru leverages TensorFlow Lite (TFLite) to enable real-time plant disease diagnosis using mobile devices, even in offline settings—a critical feature for remote farming areas. Future developments aim to incorporate TinyML and TensorFlow Lite Micro (TFLM) into embedded sensors deployed across remote farms to improve data collection, analysis, and agricultural decision-making [72].

Beyond agriculture, TinyML is also poised to make a transformative impact in environmental monitoring. It enables the deployment of intelligent, robust devices that monitor critical environmental parameters such as air quality, water purity, and climate conditions. These devices are built to operate under harsh conditions and are resilient to noise and shifts in data distribution. Their low-power design allows operation in remote or infrastructure-poor environments. The widespread deployment of TinyML-enabled devices supports real-time data availability, enabling the early detection and resolution of environmental issues. Smart farming initiatives, driven by the convergence of TinyML and intelligent sensing technologies, are ushering in a new era of agricultural and environmental innovation.

10.4.3.1   Crop Management

Crop management involves the application of diverse techniques aimed at cultivating and maintaining crops in an efficient and sustainable manner. These techniques include optimized irrigation and fertilization practices, effective pest and disease control, and strategies to maximize crop yield. TinyML offers a promising toolset in this domain by enabling on-device data collection and real-time analysis of environmental variables such as soil moisture and temperature. This allows for actionable insights and predictive analytics that enhance both productivity and resource efficiency. In recent years, researchers have introduced numerous TinyML-based innovations to assist farmers and agricultural scientists in advancing crop management practices. A few notable examples are outlined below.

One implementation involves an embedded machine learning pipeline designed for use in laboratories, greenhouses, farms, or gardens [73]. This pipeline guides users through the process of developing a TinyML model, beginning with data collection based on best practices for monitoring plant parameters. A convolutional neural network (CNN) is then trained to estimate the leaf area index (LAI) and predict the plant's growth stage. After training, the model is compressed and converted to TensorFlow Lite (TFLite) format for deployment on microcontroller units (MCUs). The Sony Spresense board was used as the target hardware for testing and validation.

Another example is a TinyML-based drought stress detection system for soybeans, which utilizes a Raspberry Pi Zero W coupled with a Sony IMX219 camera module [74]. The system captures real-time images of soybean crops and uses a CNN model to detect visual signs of drought stress. The model, originally developed using conventional deep learning frameworks, is converted into TFLite format for execution on the Raspberry Pi. Detected anomalies are transmitted to a web platform for visualization and further analysis.

A third application focuses on the detection of grapevine leaf esca disease using a compressed CNN model optimized for low-power, real-time inference [75]. Several compression techniques were explored, including canonical polyadic (CP) decomposition, Tucker decomposition, and truncated Singular value decomposition (tSVD), to identify the best trade-off between model size and accuracy. CP decomposition was ultimately selected for compressing the CNN model. Post-training quantization was then applied using TFLite to produce an 8-bit model suitable for deployment on an OpenMV Cam STM32H7. This compact vision system was mounted on an agricultural vehicle traveling at a constant speed through vineyard rows, performing disease detection in situ.

10.4.3.2   Plant Disease

Recent advancements in plant disease detection have increasingly favored image processing-based deep learning approaches over traditional RNA analysis methods. This shift aims to enable earlier and more accurate diagnosis of plant diseases, thereby helping to mitigate significant economic losses for farmers. The dominant framework in this domain employs multilayer convolutional neural network (CNN) architectures. In the initial phase, the CNN automatically extracts and learns hierarchical features from labeled training images. These extracted features are then fed into an artificial neural network (ANN) in the subsequent phase, where they are processed using learned weights, bias functions, and non-linear activation functions. This two-stage architecture has proven effective in achieving high classification accuracy in plant disease identification tasks [76].

10.4.3.3 Agricultural Advisory System

Several chatbot-driven systems have been developed to assist farmers by delivering personalized and context-specific responses to their queries [77]. These systems leverage online resources, such as agricultural documents and knowledge bases, as training data. Using natural language processing (NLP) techniques [78], a deep learning-based recurrent neural network (RNN) framework is constructed to understand and generate human-like responses. Farmers can interact with these systems by asking a wide range of questions ranging from crop suitability in specific geographical regions to the appropriate use of pesticides and fertilizers enabling timely access to critical agricultural information.

10.4.3.4 Smart Irrigation

An advanced agro-environmental management system has been developed that combines moisture sensors with real-time video analysis of soil images to optimize irrigation practices [79]. The system employs the VGG-19 deep learning model to classify soil image types and determine the appropriate amount of irrigation water, taking into account the specific crop requirements. In a separate low-cost solution, a microcontroller-based device built on the ESP32-CAM platform is proposed for monitoring central pivot irrigation systems [80]. This device uses an onboard camera to capture data from digital water meters and processes the images using a TinyML model deployed on the MCU. The processed results are then transmitted to a central server using LoRaWAN, facilitating efficient remote monitoring and data management.

10.4.3.5 Smart Greenhouse

A greenhouse is a specialized structure designed to create a controlled environment that optimizes conditions for plant growth. Key environmental variables such as temperature, humidity, light intensity, and nutrient levels can be precisely regulated, enabling the year-round cultivation of crops, including fruits, vegetables, and ornamental plants, independent of external climatic conditions.

Within greenhouse environments, TinyML offers significant potential to improve crop productivity by enabling on-device machine learning for real-time data collection, environmental monitoring, and predictive analytics, all on resource-constrained hardware.

One notable implementation involves a multi-label TinyML model based on a multilayer perceptron (MLP) architecture, designed for microclimate management in strawberry greenhouses [81]. This model supports fine-grained control over multiple environmental variables simultaneously, promoting optimal crop conditions.

Another innovative approach features a collaborative system that integrates IoT devices with edge nodes to manage greenhouse irrigation. Each IoT node is equipped with an embedded TinyML model based on a decision tree, which independently evaluates plant needs based on local sensor data. These decisions are then communicated to an edge node, which aggregates inputs from all sensors associated with a specific sprinkler. The edge node employs a higher-level machine learning model to determine the appropriate response such as "no action," "irrigation," or "fertigation," based on the collective sensor feedback [82].

10.5   CONCLUSION

This chapter provides a comprehensive exploration of TinyML, highlighting its emerging applications, technical challenges, and future directions. By enabling the deployment of machine learning models on resource-constrained embedded devices, TinyML unlocks innovative solutions across diverse fields such as smart vehicles, anomaly detection, smart farming, and beyond. Its advantages, including reduced latency, energy efficiency, and enhanced privacy, position it as a key technology in the future of edge computing.

However, the development and deployment of TinyML systems face several challenges, such as optimizing models for limited resources, managing real-time data, and the lack of standardized tools for performance evaluation and benchmarking. Proposed solutions, including hardware-aware compilers, AutoML automation, and event-driven processing, offer promising pathways to address these obstacles.

Future research should focus on refining agile development methodologies, fostering interdisciplinary expertise, and establishing standardized benchmarks to facilitate broader adoption of TinyML. Additionally, domains like healthcare, environmental monitoring, and smart cities stand to benefit significantly from TinyML advancements, particularly in regions with limited traditional infrastructure.

In summary, TinyML represents a transformative technological shift, merging artificial intelligence with embedded systems to reshape our interaction with the digital and physical world. As the field continues to evolve, TinyML is poised to expand its applications and societal impact, solidifying its role as a cornerstone of technological innovation in the years to come.

REFERENCES

1. W. Shi, J. Cao, Q. Zhang, Y. Li, and L. Xu, "Edge computing: Vision and challenges," *IEEE Internet of Things Journal*, vol. 3, no. 5, pp. 637–646, Oct. 2016.
2. R. Oshana and M. Kraeling, *Software Engineering for Embedded Systems: Methods, Practical Techniques, and Applications*. Newnes, 2019.
3. I. Lamaakal, I. Ouahbi, K. El Makkaoui, Y. Maleh, P. Pławiak, and F. Alblehai, "A TinyDL model for gesture-based air handwriting Arabic numbers and simple Arabic letters recognition," *IEEE Access*, vol. 12, pp. 76589–76605, 2024.
4. I. Lamaakal, N. El Mourabit, K. El Makkaoui, I. Ouahbi and Y. Maleh, "Efficient gesture-based recognition of tifinagh characters in air handwriting with a TinyDL model," *2024 Sixth International Conference on Intelligent Computing in Data Sciences (ICDS), Marrakech, Morocco*, 2024, pp. 1–8, doi: 10.1109/ICDS62089.2024.10756483.
5. I. Lamaakal, Y. Maleh, I. Ouahbi, K. El Makkaoui, and A. A. Abd El-Latif, "A deep learning-powered TinyML model for gesture-based air handwriting simple arabic letters recognition," in *International Conference on Digital Technologies and Applications*, Benguerir, Morocco. Cham: Springer Nature Switzerland, May 2024, pp. 32–42.

6. I. Lamaakal, S. Essahraui, Y. Maleh, K. El Makkaoui, I. Ouahbi, M. F. Bouami, A. A. Abd El-Latif, M. Almousa, J. Peng, and D. Niyato, "A Comprehensive Survey on Tiny Machine Learning for Human Behavior Analysis," *IEEE Internet of Things Journal*, vol. 12, pp. 32419–32443, 2025. doi: 10.1109/JIOT.2025.3565688.

7. S. Amershi, A. Begel, C. Bird, R. DeLine, H. Gall, E. Kamar, N. Nagappan, B. Nushi, and T. Zimmermann, "Software engineering for machine learning: A case study," [Online]. Available: https://www.microsoft.com/enus/research/uploads/prod/2019/03/amershiicse-2019_Software_Engineering_for_Machine_Learning.pdf

8. G. Giray, "A software engineering perspective on engineering machine learning systems: State of the art and challenges," *Journal of Systems and Software*, vol. 180, p. 111031, 2021.

9. C. R. Banbury, V. J. Reddi, M. Lam, W. Fu, A. Fazel, J. Holleman, X. Huang, R. Hurtado, D. Kanter, A. Lokhmotov, and D. Patterson, "Benchmarking TinyML systems: Challenges and direction," *arXiv preprint arXiv:2003.04821*, 2021. [Online]. Available: https://doi.org/10.48550/arXiv.2003.04821

10. D. Doria, I. Ernst, and B. Kadlec, "CVPR18: Tutorial: Software Engineering in Computer Vision Systems," *Computer Vision Foundation*, Jun. 2018.

11. R. P. Adams, M. Mattina, and P. N. Whatmough, "SpArSe: Sparse architecture search for CNNs on resource-constrained microcontrollers," arXiv preprint arXiv:1905.12107, 2019.

12. F. Paissan, A. Ancilotto, and E. Farella, "PhiNets: A scalable backbone for low-power AI at the edge," *ACM Transactions on Embedded Computing Systems*, vol. 21, pp. 1–18, 2021, doi: 10.1145/3510832.

13. R. David, J. Duke, A. Jain, V. Janapa Reddi, N. Jeffries, J. Li, N. Kreeger, I. Nappier, M. Natraj, T. Wang, and P. Warden, "TensorFlow lite micro: Embedded machine learning on TinyML systems," arXiv preprint arXiv:2010.08678, 2021.

14. C. Sironi, "TinyMLtalks: Machine learning for event cameras," *tinyML Foundation*, Oct. 2021. [Online]. Available: https://www.tinyml.org/event/tinyml-talks-machine-learning-for-event-cameras/

15. M. Shafique, T. Theocharides, V. Reddy, and B. Murmann, "TinyML: Current progress, research challenges, and future roadmap," *Proceedings of the 58th ACM/IEEE Design Automation Conf. (DAC)*, Dec. 2021, pp. 1303–1306, doi: 10.1109/DAC18074.2021.9586232.

16. M. Gielda, "Running TF Lite on microcontrollers without hardware in Renode," *tinyML Foundation*, Feb. 2022. [Online]. Available: https://cms.tinyml.org/wpcontent/uploads/talks2020/

17. A. Lomuscio, "TinyML Talks:Verification of ML-based AI systems and its applicability in edge ML," *tinyML Foundation*, Oct. 2021. [Online]. Available:https://cms.tinyml.org/wpcontent/uploads/talks2021/tinyML_Talks_Alessio_Lomuscio_211005.pdf

18. C. R. Banbury, V. J. Reddi, P. Torelli, J. Holleman, N. Jeffries, C. Kiraly, P. Montino, D. Kanter, S. Ahmed, D. Pau, and U. Thakker, "MLPerf Tiny Benchmark," in *Proceedings of the 2021 ACM International Conference on Compilers, Architectures, and Synthesis for Embedded Systems (CASES)*, 2021, doi: 10.1145/3528227.3528569.

19. M. Loukides, "TinyML: The challenges and opportunities of low-power ML applications," *O'Reilly Radar*, Oct. 2019. [Online]. Available: https://www.oreilly.com/radar/tinyml-the-challenges-and-opportunities-of-low-power-ml-applications/

20. M. Shafique, M. Naseer, T. Theocharides, C. Kyrkou, O. Mutlu, L. Orosa, and J. Choi, "Robust machine learning systems: Challenges, current trends, perspectives, and the road ahead," *IEEE Design & Test*, vol. 37, no. 2, pp. 30–57, Apr. 2020, doi: 10.1109/MDAT.2020.2971774.

21. H. Cai, C. Gan, L. Zhu, and S. Han, "Tiny transfer learning: Towards memory-efficient on-device learning," *CoRR*, arXiv:2007.11622v1, 2020.

22. L. Lai, N. Suda, and V. Chandra, "CMSIS-NN: Efficient neural network kernels for Arm Cortex-M CPUs," *CoRR*, arXiv:1801.06601, 2018.

23. D. Xu, T. Li, Y. Li, X. Su, S. Tarkoma, T. Jiang, J. Crowcroft, and P. Hui, "Edge intelligence: Architectures, challenges, and applications," *CoRR*, arXiv:2003.12172, 2020.

24. C. Banbury, C. Zhou, I. Fedorov, R. M. Navarro, U. Thakker, D. Gope, V. J. Reddi, M. Mattina, and P. N. Whatmough, "MicroNets: Neural network architectures for deploying TinyML applications on commodity microcontrollers," *CoRR*, arXiv:2010.11267, 2021.

25. A. Makhshari and A. Mesbah, "IoT bugs and development challenges," in *Proceedings of the 2021 IEEE/ACM 43rd International Conference on Software Engineering (ICSE)*, Madrid, ES, pp. 460–472, 2021.

26. M. J. Islam, G. Nguyen, R. Pan, and H. Rajan, "A comprehensive study on deep learning bug characteristics," in *Proceedings of the 2019 ACM Joint Meeting on European Software Engineering Conference and Symposium on the Foundations of Software Engineering (ESEC/FSE)*, Tallinn, Estonia, 2019, doi: 10.1145/3338906.3338955.

27. R. Sanchez-Iborra and A. F. Skarmeta, "TinyML-enabled frugal smart objects: challenges and opportunities," *IEEE Circuits and Systems Magazine*, vol. 20, no. 3, pp. 4–18, 2020.

28. Z. Wu, M. Jiang, H. Li, and X. Zhang, "Mapping the knowledge domain of smart city development to urban sustainability: a scientometric study," *Journal of Urban Technology*, vol. 28, no. 1, pp. 29–53, 2021.

29. I. Fedorov et al., "TinyLSTMs: Efficient neural speech enhancement for hearing aids," in *Proceedings of the Interspeech 2020*, Shanghai, China, Oct. 25–29, 2020.

30. D. Rossi et al., "PULP: A parallel ultra low power platform for next generation IoT applications," in *Proceedings of the 2015 IEEE Hot Chips 27 Symposium (HCS)*, Cupertino, CA, USA, Aug. 22–25, 2015, pp. 1–39.

31. M. Monfort Grau, *TinyML: From Basic to Advanced Applications*, Bachelor's thesis, Universitat Politècnica de Catalunya, Barcelona, Spain, 2021.

32. U. S. Shanthamallu and A. Spanias, "Machine and deep learning applications," in *Machine and Deep Learning Algorithms and Applications*, Springer, Berlin/Heidelberg, Germany, 2022, pp. 59–72.

33. L. Atzori, A. Iera, and G. Morabito, "The internet of things: A survey," *Computer Networks*, vol. 54, no. 15, pp. 2787–2805, Oct. 2010.

34. X. Cheng, C. Chen, W. Zhang, and Y. Yang, "5G-enabled cooperative intelligent vehicular (5GenCIV) framework: When Benz meets Marconi," *IEEE Intelligent Systems*, vol. 32, no. 3, pp. 53–59, May–Jun. 2017.

35. L. Liang, H. Peng, G. Y. Li, and X. Shen, "Vehicular communications: A network layer perspective," *IEEE Transactions on Vehicular Technology*, vol. 66, no. 12, pp. 10647–10659, Dec. 2017.

36. H. Ye, G. Y. Li, and B.-H. F. Juang, "Deep reinforcement learning based resource allocation for V2V communications," *IEEE Transactions on Vehicular Technology*, vol. 68, no. 4, pp. 3163–3173, Apr. 2019.

37. "A look at the future of 5G," *IEEE Spectrum*. [Online]. Available: https://spectrum.ieee.org/computing/software/a-look-at-the-future-of-5g [Accessed: Nov. 28, 2023].

38. K. Hong, D. Lillethun, U. Ramachandran, B. Ottenwälder, and B. Koldehofe, "Mobile fog: A programming model for large-scale applications on the Internet of Things," in *Proceedings of the ACM SIGCOMM Workshop on Mobile Cloud Computing*, Hong Kong, China, Aug. 2013, pp. 15–20.

39. S. Roy, R. Bose, and D. Sarddar, "A fog-based DSS model for driving rule violation monitoring framework on the internet of things," *International Journal of Advanced Science and Technology*, vol. 82, pp. 23–32, 2015.

40. G. Vashitz, D. Shinar, and Y. Blum, "In-vehicle information systems to improve traffic safety in road tunnels," *Transportation Research Part F: Traffic Psychology and Behaviour*, vol. 11, no. 1, pp. 61–74, 2008.
41. S. J. Miah and R. Ahamed, "A cloud-based DSS model for driver safety and monitoring on Australian roads," *International Journal of Emerging Sciences*, vol. 1, no. 4, p. 634, 2011.
42. H. Vahdat-Nejad, A. Ramazani, T. Mohammadi, and W. Mansoor, "A survey on context aware vehicular network applications," *Vehicular Communications*, vol. 3, pp. 43–57, 2016.
43. S. B. Chougule, B. S. Chaudhari, S. N. Ghorpade, and M. Zennaro, "Exploring computing paradigms for electric vehicles: From cloud to edge intelligence, challenges and future directions," *World Electric Vehicle Journal*, vol. 15, no. 2, p. 39, 2024, doi: 10.3390/wevj15020039.
44. M. Baldauf, S. Dustdar, and F. Rosenberg, "A survey on context aware systems," *International Journal of Ad Hoc and Ubiquitous Computing.*, vol. 2, pp. 263–277, 2007.
45. T. E. Bogale, X. Wang, and L. B. Le, "Machine intelligence techniques for next-generation context-aware wireless networks," *Computer Science Inforamtion Theory*, arXiv:1801.04223, 2018.
46. Q. He, J. Liu, C. Wang, and B. Li, "Coping with heterogeneous video contributors and viewers in crowdsourced live streaming: A cloud-based approach," *IEEE Transactions on Multimedia*, vol. 18, pp. 916–928, 2016.
47. G. Zhuo, Q. Jia, L. Guo, M. Li, and P. Li, "Privacy-preserving verifiable set operation in big data for cloud-assisted mobile crowdsourcing," *IEEE Internet of Things Journal*, vol. 4, pp. 572–582, 2017.
48. C. Huang and K. Xu, "Reliable real-time streaming in vehicular cloud-fog computing networks," in *Proceedings of the IEEE Conference on Communications in China*, Chengdu, China, Jul. 2016, pp. 1–6.
49. G. Grassi, P. Bahl, K. Jamieson, and G. Pau, "Park Master: An in-vehicle, edge-based video analytics service for detecting open parking spaces in urban environments," in *Proceedings of the ACM/IEEE Symposium on Edge Computing*, San Jose, CA, USA, Apr. 2017, p. 16.
50. S. N. Ghorpade, M. Zennaro, and B. S. Chaudhari, "GWO model for optimal localization of IoT-enabled sensor nodes in smart parking systems," *IEEE Transactions on Intelligent Transportation Systems*, vol. 22, pp. 1217–1224, 2021.
51. H. S. Park, M. W. Park, K. H. Won, K.-H. Kim, and S. K. Jung, "In-vehicle AR-HUD system to provide driving-safety information," *ETRI Journal*, vol. 35, pp. 1038–1047, 2013.
52. J. Manokaran and G. Vairavel, "Smart anomaly detection using data-driven techniques in IoT edge: A survey," in *Proceedings of Third International Conference on Communication, Computing and Electronics Systems,* Springer, Cham, Switzerland, 2022, pp. 685–702.
53. A. Chatterjee and B. S. Ahmed, "IoT anomaly detection methods and applications: A survey," *IEEE Internet of Things Journal*, vol. 19, Aug. 2022, Art. no. 100568.
54. I. Lamaakal, K. El Makkaoui, I. Ouahbi, and Y. Maleh, "A TinyML model for gesture-based air handwriting Arabic numbers recognition," *Procedia Computer Science*, vol. 236, pp. 589–596, 2024.
55. I. Lamaakal, Z. Charroud, Y. Maleh, I. Ouahbi, and K. E. Makkaoui, "Optimizing breast calcification detection in mammography using PySpark: A big data and machine learning approach," in *Proceedings of the International Conference on Data Analytical Management*, Singapore: Springer, 2025, pp. 617–627.

56. B. Nassif, M. A. Talib, Q. Nasir, and F. M. Dakalbab, "Machine learning for anomaly detection: A systematic review," *IEEE Access*, vol. 9, pp. 78658–78700, 2021.

57. A. Xenakis, A. Karageorgos, E. Lallas, A. E. Chis, and H. González-Vélez, "Towards distributed IoT/Cloud-based fault detection and maintenance in industrial automation," *Procedia Computer Science Journal*, vol. 151, pp. 683–690, Jan. 2019.

58. A. Mostafavi and A. Sadighi, "A novel online machine learning approach for real-time condition monitoring of rotating machines," in *Proceedings of the 9th RSI International Conference on Robotics and Mechatronics (ICRoM)*, Tehran, Iran, Islamic Republic of, Nov. 2021, pp. 267–273.

59. M. Antonini, M. Pincheira, M. Vecchio, and F. Antonelli, "A TinyML approach to non-repudiable anomaly detection in extreme industrial environments," in *Proceedings of the IEEE Internationl Workshop Metrology Industry 4.0 & IoT (MetroInd4.0&IoT)*, Trento, Italy, Jun. 2022, pp. 397–402.

60. M. Lord and A. Kaplan, "Mechanical anomaly detection on an embedded microcontroller," in *Proceedings of the International Conference on Computational Science and Computational Intelligence (CSCI)*, Las Vegas, NV, USA, Dec. 2021, pp. 562–568.

61. V. M. Oliveira and A. H. Moreira, "Edge AI system using a thermal camera for industrial anomaly detection," in *Proceedings of the International Summit Smart City 360°*, Springer, Cham, Switzerland, 2022, pp. 172–187.

62. D. Pau, A. Khiari, and D. Denaro, "Online learning on tiny microcontrollers for anomaly detection in water distribution systems," in *Proceedings of the IEEE 11th IEEE International Conference on Consumer Electronics (ICCE-Berlin)*, Berlin, Germany, Nov. 2021, pp. 1–6.

63. M. Cardoni, D. P. Pau, L. Falaschetti, C. Turchetti, and M. Lattuada, "Online learning of oil leak anomalies in wind turbines with block-based binary reservoir," *Electronics*, vol. 10, no. 22, Nov. 2021, Art. no. 2836.

64. P. Andrade, I. Silva, G. Signoretti, M. Silva, J. Dias, L. Marques, and D. G. Costa, "An unsupervised TinyML approach applied for pavement anomalies detection under the Internet of intelligent vehicles," in *Proceedings of the IEEE Internationl on Workshop Metrology for Industry 4.0 & IoT (MetroInd4.0&IoT)*, Rome, Italy, Jun. 2021, pp. 642–647.

65. A. Mitra, S. L. T. Vangipuram, A. K. Bapatla, V. K. V. V. Bathalapalli, S. P. Mohanty, E. Kougianos, and C. Ray, "Everything you wanted to know about smart agriculture," *arXiv preprint arXiv:2201.04754*, 2022.

66. S. Condran, M. Bewong, M. Z. Islam, L. Maphosa, and L. Zheng, "Machine learning in precision agriculture: A survey on trends, applications and evaluations over two decades," *IEEE Access*, vol. 10, pp. 73786–73803, 2022.

67. S. N. Ghorpade, M. Zennaro, and B. S. Chaudhari, "Towards green computing: Intelligent bioinspired agent for IoT-enabled wireless sensor networks," *International Journal of Sensor Networks*, vol. 35, no. 2, pp. 121–131, Feb. 2021, doi: 10.1504/IJSNET.2021.113632.

68. G. Singh, A. Singh, and G. Kaur, "Role of artificial intelligence and the Internet of Things in agriculture," in *Artificial Intelligence to Solve Pervasive Internet of Things Issues*, Elsevier, Amsterdam, The Netherlands, 2021, pp. 317–330.

69. S. O. Ooko, M. M. Ogore, J. Nsenga, and M. Zennaro, "TinyML in Africa: Opportunities and challenges," in *Proceedings of the IEEE Globecom Workshops (GC Wkshps)*, Madrid, Spain, Dec. 2021, pp. 1–6.

70. V. K. Quy, N. V. Hau, D. V. Anh, N. M. Quy, N. T. Ban, S. Lanza, G. Randazzo, and A. Muzirafuti, "IoT-enabled smart agriculture: Architecture, applications, and challenges," *Applied Science*, vol. 12, no. 7, Mar. 2022, Art. no. 3396.

71. PlantVillage. [Online]. Available: https://plantvillage.psu.edu/. [Accessed: Feb. 5, 2023].

72. N. Schizas, A. Karras, C. Karras, and S. Sioutas, "TinyML for ultra-low power AI and large-scale IoT deployments: A systematic review," *Future Internet*, vol. 14, no. 12, Dec. 2022, Art. no. 363.

73. D. Sheth, B. Sudharsan, J. G. Breslin, and M. I. Ali, "Embedded ML pipeline for precision agriculture," in *Proceedings of the 21st ACM/IEEE nternational Conference on Information Processing in Sensor Networks (IPSN)*, Milano, Italy, May 2022, pp. 527–528.

74. P. Ramos-Giraldo, S. C. Reberg-Horton, S. Mirsky, E. Lobaton, A. M. Locke, E. Henriquez, A. Zuniga, and A. Minin, "Low-cost smart camera system for water stress detection in crops," in *Proceedings of the IEEE SENSORS*, Rotterdam, The Netherlands, Dec. 2020, pp. 1–4.

75. L. Falaschetti, L. Manoni, R. C. F. Rivera, D. Pau, G. Romanazzi, O. Silvestroni, V. Tomaselli, and C. Turchetti, "A low-cost, low-power and real-time image detector for grape leaf esca disease based on a compressed CNN," *IEEE Journal on Emerging and Selected Topics in Circuits and Systems*, vol. 11, no. 3, pp. 468–481, Sep. 2021.

76. N. Gobalakrishnan, K. Pradeep, C. J. Raman, L. J. Ali, and M. P. Gopinath, "A systematic review on image processing and machine learning techniques for detecting plant diseases," in *Proceedings of the International Conference on Communication and Signal Processing (ICCSP)*, Chennai, India, Jul. 2020, pp. 0465–0468, doi: 10.1109/ICCSP48568.2020.9182046.

77. P. Y. Niranjan, V. S. Rajpurohit, and R. Malgi, "A survey on chat-bot system for agriculture domain," in *Proceedings of the 1st International Conference on Advances in Information Technology (ICAIT)*, Chikmagalur, India, Jul. 2019, pp. 99–103, doi: 10.1109/ICAIT47043.2019.8987429.

78. I. Lamaakal, Y. Maleh, K. El Makkaoui, I. Ouahbi, P. Pławiak, O. Alfarraj, and A. A. Abd El-Latif, "Tiny language models for automation and control: Overview, potential applications, and future research directions," *Sensors*, vol. 25, no. 5, p. 1318, 2025.

79. S. W. Mohammed, N. R. Soora, N. Polala, and S. Saman, "Smart water resource management by analyzing the soil structure and moisture using deep learning," in *IoT with Smart Systems*, J. Choudrie, P. Mahalle, T. Perumal, and A. Joshi, Eds. Singapore: Springer Nature, 2023, pp. 709–719.

80. D. M. Matilla, A. L. Murciego, D. M. Jiménez-Bravo, A. S. Mendes, and V. R. Leithardt, "Low cost edge computing devices and novel user interfaces for monitoring pivot irrigation systems based on Internet of Things and LoRaWAN technologies," *Biosystems Engineering*, vol. 223, pp. 14–29, Nov. 2022.

81. I. Ihoume, R. Tadili, N. Arbaoui, M. Benchrifa, A. Idrissi, and M. Daoudi, "Developing a multi-label TinyML machine learning model for an active and optimized greenhouse microclimate control from multivariate sensed data," *Artificial Intelligence in Agriculture*, vol. 6, pp. 129– 137, Jan. 2022.

82. R. Sanchez-Iborra, A. Zoubir, A. Hamdouchi, A. Idri, and A. Skarmeta, "Intelligent and efficient IoT through the cooperation of TinyML and edge computing," *Informatica*, vol. 34, no. 1, pp. 147–168, 2023.

Index

For Product Safety Concerns and Information please contact our EU
representative GPSR@taylorandfrancis.com
Taylor & Francis Verlag GmbH, Kaufingerstraße 24, 80331 München, Germany

www.ingramcontent.com/pod-product-compliance
Lightning Source LLC
Chambersburg PA
CBHW031950180326
41458CB00006B/1677